Multiplication

Workbook

ANSWER KEY can be found at www.preschoolprepco.com/answers

A Divison of Preschool Prep Company™

866-451-5600
www.PreschoolPrepCo.com
P.O. Box 1159, Danville CA 94526

Fill in the Answer

Fill in the blanks to complete the equations.

1 × 1 = ___	___ × 1 = 1	___ × 1 = 1
1 × 1 = ___	1 × ___ = 1	1 × ___ = 1
1 × ___ = 1	___ × 1 = 1	1 × 1 = ___
___ × 1 = 1	1 × ___ = 1	1 × 1 = ___

___	1	1	___	1	1
× 1	× 1	× ___	× 1	× ___	× 1
1	___	1	1	1	___

1	1	___	1	1	___
× 1	× ___	× 1	× ___	× 1	× 1
___	1	1	1	___	1

Activities

There is one cup.

There is one paint brush in the cup.

How many paint brushes are there? 1

Use the numbers below to make equations.

1 1 1

1	×	1	=	1
	×	1	=	
1	×		=	
	×	1	=	

Draw one paint brush in the cup.

_____ × _____ = _____

Write the equation to match the picture.

Solve the equations to get the girl to the easel.

Fill in the Answer

Fill in the blanks to complete the equations.

$\square \times 1 = 2$ $\quad$ $2 \times 1 = \square$ $\quad$ $2 \times 1 = \square$

$1 \times 2 = \square$ $\quad$ $\square \times 2 = 2$ $\quad$ $\square \times 1 = 2$

$2 \times \square = 2$ $\quad$ $1 \times \square = 2$ $\quad$ $2 \times \square = 2$

$1 \times 2 = \square$ $\quad$ $\square \times 2 = 2$ $\quad$ $1 \times \square = 2$

$$\begin{array}{r}1\\ \times\ 2\\ \hline \square\end{array}\quad\begin{array}{r}\square\\ \times\ 1\\ \hline 2\end{array}\quad\begin{array}{r}1\\ \times\ \square\\ \hline 2\end{array}\quad\begin{array}{r}\square\\ \times\ 1\\ \hline 2\end{array}\quad\begin{array}{r}1\\ \times\ \square\\ \hline 2\end{array}\quad\begin{array}{r}1\\ \times\ 2\\ \hline \square\end{array}$$

$$\begin{array}{r}1\\ \times\ \square\\ \hline 2\end{array}\quad\begin{array}{r}1\\ \times\ 2\\ \hline \square\end{array}\quad\begin{array}{r}\square\\ \times\ 1\\ \hline 2\end{array}\quad\begin{array}{r}1\\ \times\ \square\\ \hline 2\end{array}\quad\begin{array}{r}2\\ \times\ 1\\ \hline \square\end{array}\quad\begin{array}{r}\square\\ \times\ 1\\ \hline 2\end{array}$$

Activities

There are two cups.

There is one toothbrush in each cup.

How many toothbrushes are there in all? ______

Use the numbers below to make equations.

1 2 2

___ × 2 = ___

2 × ___ = ___

___ × 1 = ___

1 × ___ = ___

Draw two tooth brushes in the cup.

______ × ______ = ______

Write the equation to match the picture.

Solve the problems to get the toothbrush to the dinosaur.

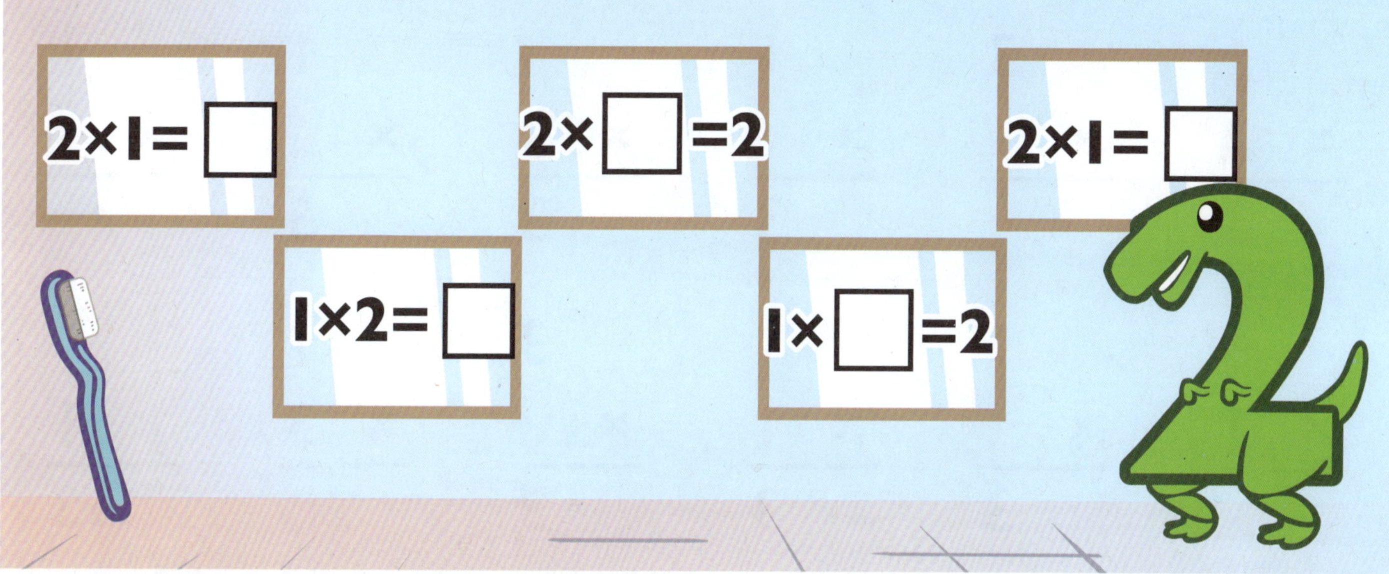

Fill in the Answer

Fill in the blanks to complete the equations.

$\square \times 3 = 3$	$3 \times 1 = \square$	$\square \times 3 = 3$
$3 \times 1 = \square$	$3 \times \square = 3$	$3 \times 1 = \square$
$1 \times \square = 3$	$\square \times 1 = 3$	$\square \times 3 = 3$
$3 \times \square = 3$	$1 \times 3 = \square$	$1 \times \square = 3$

$$\begin{array}{r}\square\\ \times\ 3\\ \hline 3\end{array}\qquad \begin{array}{r}3\\ \times\ 1\\ \hline \square\end{array}\qquad \begin{array}{r}1\\ \times\ \square\\ \hline 3\end{array}\qquad \begin{array}{r}\square\\ \times\ 3\\ \hline 3\end{array}\qquad \begin{array}{r}1\\ \times\ \square\\ \hline 3\end{array}\qquad \begin{array}{r}1\\ \times\ 3\\ \hline \square\end{array}$$

$$\begin{array}{r}3\\ \times\ 1\\ \hline \square\end{array}\qquad \begin{array}{r}1\\ \times\ \square\\ \hline 3\end{array}\qquad \begin{array}{r}\square\\ \times\ 1\\ \hline 3\end{array}\qquad \begin{array}{r}3\\ \times\ \square\\ \hline 3\end{array}\qquad \begin{array}{r}1\\ \times\ 3\\ \hline \square\end{array}\qquad \begin{array}{r}\square\\ \times\ 1\\ \hline 3\end{array}$$

Activities

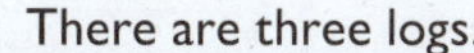

There is one water stream.

There are three logs.

How many logs are in the stream?______

Use the numbers below to make equations.

1 3 3

1	×		=	
	×		=	3
3	×		=	
	×	3	=	

Draw one log on each stream.

_____ × _____ = _____

Write the equation to match the picture.

Solve the problems to get the beaver to the stream.

3×1= ☐

1×3= ☐

3×1= ☐

☐ ×3= ☐

Fill in the Answer

Fill in the blanks to complete the equations.

$1 \times \square = 4$ $1 \times 4 = \square$ $1 \times \square = 4$

$\square \times 1 = 4$ $\square \times 4 = 4$ $4 \times 1 = \square$

$4 \times 1 = \square$ $1 \times 4 = \square$ $4 \times \square = 4$

$\square \times 4 = 4$ $4 \times \square = 4$ $\square \times 1 = 4$

$\begin{array}{r} \square \\ \times\ 1 \\ \hline 4 \end{array}$ $\begin{array}{r} 4 \\ \times\ 1 \\ \hline \square \end{array}$ $\begin{array}{r} 1 \\ \times\ \square \\ \hline 4 \end{array}$ $\begin{array}{r} \square \\ \times\ 4 \\ \hline 4 \end{array}$ $\begin{array}{r} 4 \\ \times\ \square \\ \hline 4 \end{array}$ $\begin{array}{r} 1 \\ \times\ 4 \\ \hline \square \end{array}$

$\begin{array}{r} 1 \\ \times\ 4 \\ \hline \square \end{array}$ $\begin{array}{r} 1 \\ \times\ \square \\ \hline 4 \end{array}$ $\begin{array}{r} \square \\ \times\ 1 \\ \hline 4 \end{array}$ $\begin{array}{r} 1 \\ \times\ \square \\ \hline 4 \end{array}$ $\begin{array}{r} 1 \\ \times\ 4 \\ \hline \square \end{array}$ $\begin{array}{r} \square \\ \times\ 4 \\ \hline 4 \end{array}$

Activities

There is one machine.

There are four coins for the machine.

How many coins are there in all? _____

Use the numbers below to make equations.

Draw four coins on his hand.

_____ × _____ = _____

Write the equation to match the picture.

Solve the problems to get the girl to the coin machine.

Fill in the Answer

Fill in the blanks to complete the equations.

$5 \times ☐ = 5$	$1 \times 5 = ☐$	$1 \times 5 = ☐$
$☐ \times 5 = 5$	$1 \times ☐ = 5$	$5 \times ☐ = 5$
$1 \times ☐ = 5$	$☐ \times 1 = 5$	$☐ \times 5 = 5$
$☐ \times 1 = 5$	$5 \times 1 = ☐$	$1 \times 5 = ☐$

☐	1	5	☐	1	1
× 5	× 5	× ☐	× 1	× ☐	× 5
5	☐	5	5	5	☐

5	1	☐	5	5	☐
× 1	× ☐	× 1	× ☐	× 1	× 5
☐	5	5	5	☐	5

Activities

There are five kids.

Each kid gets one ice cream bar.

How many bars are there in all? _____

Use the numbers below to make equations.

Draw one scoop of ice cream in each cone.

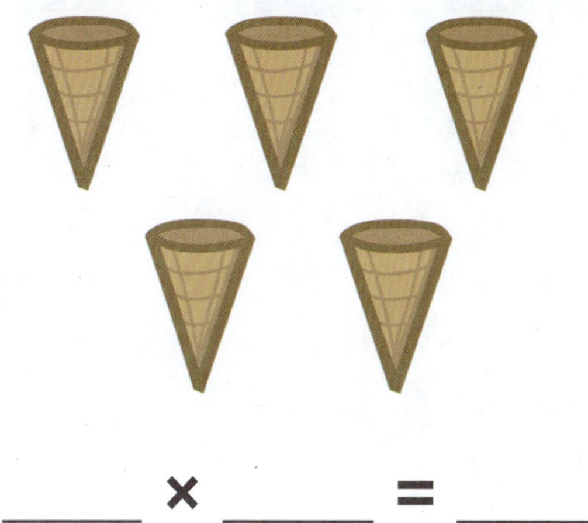

_____ × _____ = _____

Write the equation to match the picture.

Fill in the Answer

Fill in the blanks to complete the equations.

$6 \times 1 = \square$	$1 \times \square = 6$	$6 \times \square = 6$
$6 \times \square = 6$	$\square \times 1 = 6$	$1 \times 6 = \square$
$\square \times 6 = 6$	$1 \times 6 = \square$	$\square \times 6 = 6$
$6 \times 1 = \square$	$\square \times 6 = 6$	$1 \times \square = 6$

$$\begin{array}{r} \square \\ \times\ 1 \\ \hline 6 \end{array} \quad \begin{array}{r} 1 \\ \times\ 6 \\ \hline \square \end{array} \quad \begin{array}{r} 6 \\ \times\ \square \\ \hline 6 \end{array} \quad \begin{array}{r} \square \\ \times\ 6 \\ \hline 6 \end{array} \quad \begin{array}{r} 1 \\ \times\ \square \\ \hline 6 \end{array} \quad \begin{array}{r} 1 \\ \times\ 6 \\ \hline \square \end{array}$$

$$\begin{array}{r} 6 \\ \times\ 1 \\ \hline \square \end{array} \quad \begin{array}{r} 1 \\ \times\ \square \\ \hline 6 \end{array} \quad \begin{array}{r} \square \\ \times\ 1 \\ \hline 6 \end{array} \quad \begin{array}{r} 6 \\ \times\ \square \\ \hline 6 \end{array} \quad \begin{array}{r} 1 \\ \times\ 6 \\ \hline \square \end{array} \quad \begin{array}{r} \square \\ \times\ 1 \\ \hline 6 \end{array}$$

Activities

There are six ice cubes.

There is one cup.

How many cubes are in the cup? _____

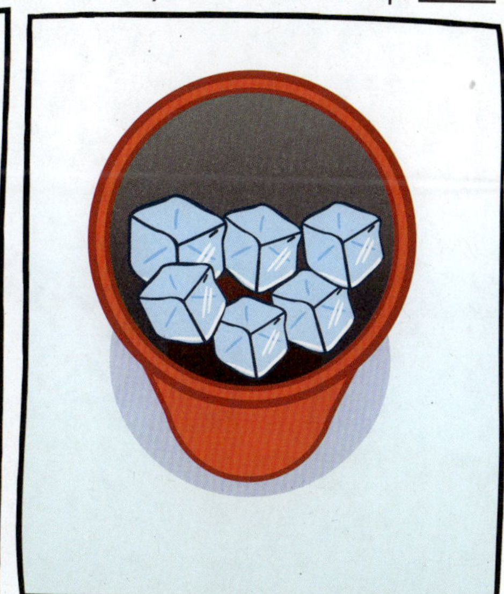

Use the numbers below to make equations.

1 6 6

Draw one fish in each fishing hole.

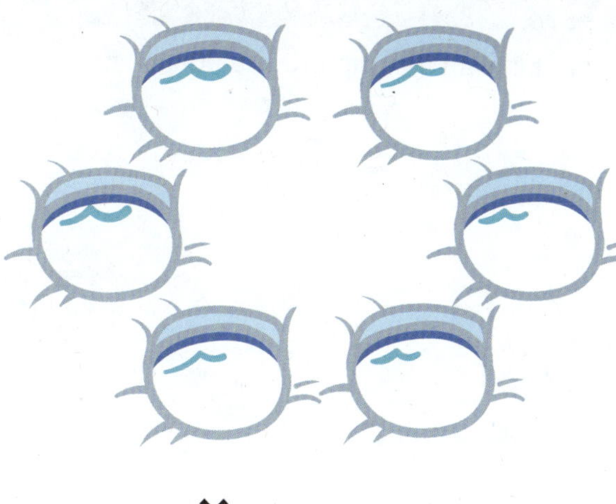

_____ × _____ = _____

Write the equation to match the picture.

Solve the problems to get the seal to the fishing hole.

Fill in the Answer

Fill in the blanks to complete the equations.

$1 \times \square = 7$ $\square \times 7 = 7$ $1 \times 7 = \square$

$7 \times 1 = \square$ $7 \times \square = 7$ $7 \times \square = 7$

$1 \times \square = 7$ $\square \times 1 = 7$ $\square \times 7 = 7$

$\square \times 1 = 7$ $7 \times 1 = \square$ $1 \times 7 = \square$

$\begin{array}{r} 1 \\ \times\ 7 \\ \hline \square \end{array}$ $\begin{array}{r} \square \\ \times\ 1 \\ \hline 7 \end{array}$ $\begin{array}{r} 1 \\ \times\ \square \\ \hline 7 \end{array}$ $\begin{array}{r} \square \\ \times\ 1 \\ \hline 7 \end{array}$ $\begin{array}{r} 1 \\ \times\ \square \\ \hline 7 \end{array}$ $\begin{array}{r} 1 \\ \times\ 7 \\ \hline \square \end{array}$

$\begin{array}{r} 1 \\ \times\ \square \\ \hline 7 \end{array}$ $\begin{array}{r} 1 \\ \times\ 7 \\ \hline \square \end{array}$ $\begin{array}{r} \square \\ \times\ 1 \\ \hline 7 \end{array}$ $\begin{array}{r} 1 \\ \times\ \square \\ \hline 7 \end{array}$ $\begin{array}{r} 7 \\ \times\ 1 \\ \hline \square \end{array}$ $\begin{array}{r} \square \\ \times\ 1 \\ \hline 7 \end{array}$

Activities

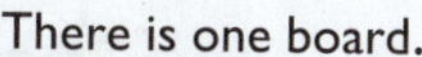

There are seven nails.

There is one board.

How many nails are in the board? ______

Use the numbers below to make equations.

Draw seven dog snacks on the dog bowl.

______ × ______ = ______

Write the equation to match the picture.

Solve the problems to get the construction worker to his hammer.

Fill in the Answer

Fill in the blanks to complete the equations.

___ × 1 = 8	1 × ___ = 8	1 × 8 = ___
8 × ___ = 8	___ × 1 = 8	8 × ___ = 8
1 × 8 = ___	8 × 1 = ___	___ × 8 = 8
___ × 8 = 8	8 × ___ = 8	8 × 1 = ___

1	___	1	___	8	1
× 8	× 1	× ___	× 8	× ___	× 8
___	8	8	8	8	___

8	1	___	1	8	___
× ___	× 8	× 8	× ___	× 1	× 1
8	___	8	8	___	8

Activities

There is one plate.

There are eight pieces of sushi.

How many pieces are there for the plate? _____

Use the numbers below to make equations.

Draw eight chocolate chips on the cookie.

_____ × _____ = _____

Write the equation to match the picture.

Solve the equations to get the panda to the bamboo.

Fill in the Answer

Fill in the blanks to complete the equations.

$1 \times \square = 9$	$1 \times 9 = \square$	$\square \times 9 = 9$
$\square \times 9 = 9$	$1 \times \square = 9$	$1 \times 9 = \square$
$9 \times 1 = \square$	$\square \times 1 = 9$	$\square \times 9 = 9$
$9 \times \square = 9$	$9 \times 1 = \square$	$9 \times \square = 9$

$\begin{array}{r} \square \\ \times\ 9 \\ \hline 9 \end{array}$	$\begin{array}{r} 1 \\ \times\ 9 \\ \hline \square \end{array}$	$\begin{array}{r} 9 \\ \times\ \square \\ \hline 9 \end{array}$	$\begin{array}{r} \square \\ \times\ 1 \\ \hline 9 \end{array}$	$\begin{array}{r} 1 \\ \times\ \square \\ \hline 9 \end{array}$	$\begin{array}{r} 1 \\ \times\ 9 \\ \hline \square \end{array}$

$\begin{array}{r} 9 \\ \times\ 1 \\ \hline \square \end{array}$	$\begin{array}{r} 1 \\ \times\ \square \\ \hline 9 \end{array}$	$\begin{array}{r} \square \\ \times\ 9 \\ \hline 9 \end{array}$	$\begin{array}{r} 9 \\ \times\ \square \\ \hline 9 \end{array}$	$\begin{array}{r} 9 \\ \times\ 1 \\ \hline \square \end{array}$	$\begin{array}{r} \square \\ \times\ 1 \\ \hline 9 \end{array}$

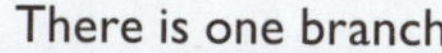

Activities

There are nine butterflies.

There is one branch.

How many butterflies are there in all? ____

Use the numbers below to make equations.

Draw one butterfly on each tree.

_____ × _____ = _____

Write the equation to match the picture.

Solve the problems to get the owl to his clock.

Fill in the Answer

Fill in the blanks to complete the equations.

1 × 10 = ___	___ × 10 = 10	10 × ___ = 10
10 × ___ = 10	10 × 1 = ___	1 × 10 = ___
___ × 1 = 10	1 × ___ = 10	___ × 10 = 10
1 × ___ = 10	___ × 1 = 10	10 × 1 = ___

___	10	1	___	10	1
× 10	× 1	× ___	× 1	× ___	× 10
10	___	10	10	10	___

1	1	___	10	10	___
× 10	× ___	× 1	× ___	× 1	× 10
___	10	10	10	___	10

Activities

There are ten slices.

There is one pepperoni on each slice.

How many pepperoni are there in all? ____

Use the numbers below to make equations.

1 10 10

Draw ten pepperonis on the pizza.

_____ × _____ = _____

Write the equation to match the picture.

Solve the equations to get the chef to his brick oven.

Fill in the Answer

Fill in the blanks to complete the equations.

I × ___ = II	II × I = ___	II × I = ___
I × II = ___	___ × II = II	___ × I = II
___ × I = II	I × ___ = II	II × ___ = II
II × ___ = II	I × II = ___	___ × II = II

I	___	II	___	I	I
× II	× I	× ___	× I	× ___	× II
___	II	II	II	II	___

II	I	___	II	I	___
× I	× ___	× II	× ___	× II	× I
___	II	II	II	___	II

Activities

There are eleven cupcakes.

Each cupcake has one candle.

How many candles are there in all? _____

Use the numbers below to make equations.

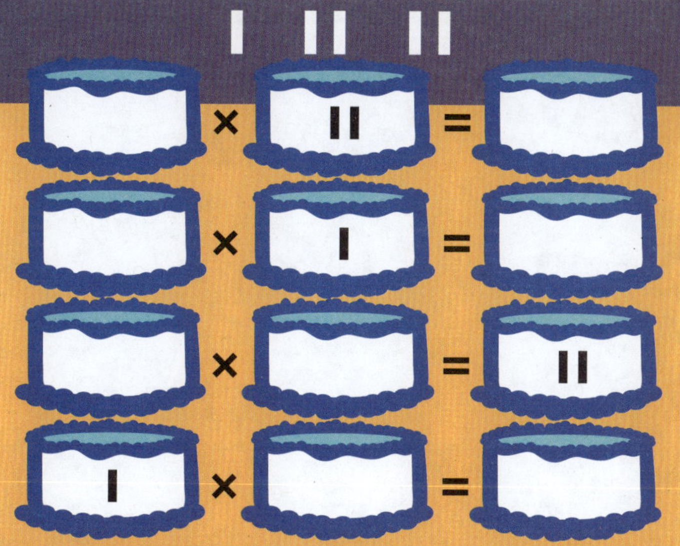

Draw one flame on each candle.

_____ × _____ = _____

Write the equation to match the picture.

Solve the problems to get the firefighter to the fire.

Fill in the Answer

Fill in the blanks to complete the equations.

1 × 12 = ____	12 × 1 = ____	____ × 12 = 12
12 × ____ = 12	____ × 12 = 12	12 × 1 = ____
____ × 1 = 12	1 × 12 = ____	____ × 1 = 12
1 × ____ = 12	1 × ____ = 12	12 × ____ = 12

1	____	12	____	1	1
× 12	× 1	× ____	× 12	× ____	× 12
____	12	12	12	12	____

12	1	____	12	1	____
× 1	× ____	× 12	× ____	× 12	× 1
____	12	12	12	____	12

Activities

There is one basket.

There are twelve balls.

How many balls are in the basket? ____

Use the numbers below to make equations.

Draw one ball in each basket.

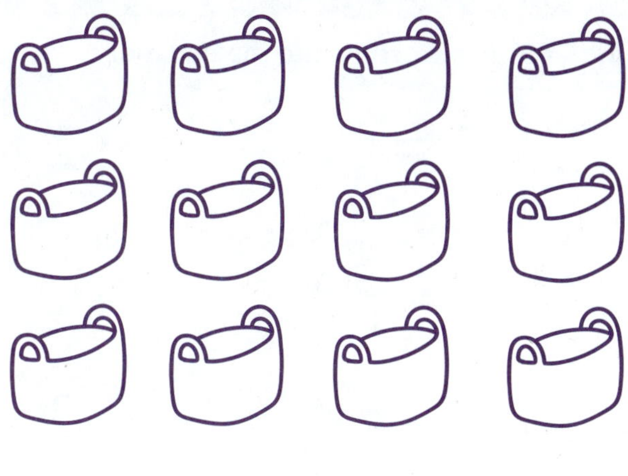

_____ × _____ = _____

Write the equation to match the picture.

Solve the equations to get the baby to his bottle.

1 × ☐ = 12

☐ × 1 = 12

1 × 12 = ☐

☐ × 1 = 12

1 × 12 = ☐

Fill in the Answer

Fill in the blanks to complete the equations.

2 × 2 = ___	2 × ___ = 4	___ × 2 = 4
___ × 2 = 4	___ × 2 = 4	2 × ___ = 4
2 × ___ = 4	2 × 2 = ___	2 × 2 = ___
2 × 2 = ___	2 × ___ = 4	___ × 2 = 4

2	___	2	___	2	2
× 2	× 2	× ___	× 2	× ___	× 2
___	4	4	4	4	___

2	2	___	2	2	___
× 2	× ___	× 2	× ___	× 2	× 2
___	4	4	4	___	4

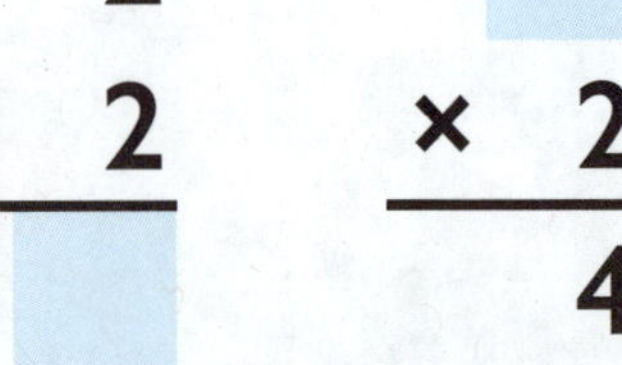

Activities

Two ears of corn grow on each stalk.

There are two stalks.

How many ears are there in all? ______

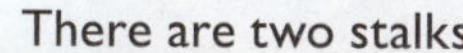

Draw two ears of corn on each stalk.

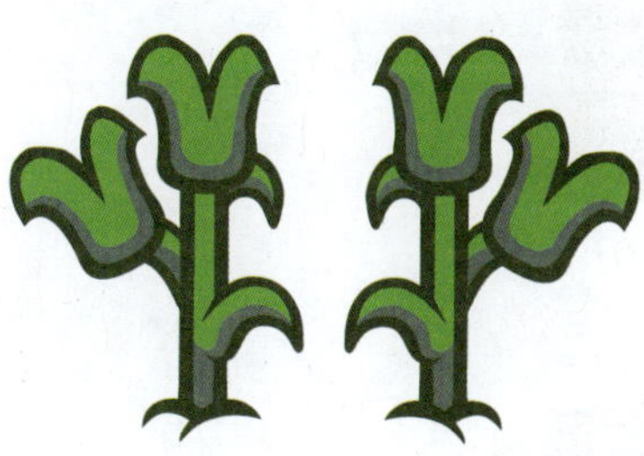

_____ × _____ = _____

Write the equation to match the picture.

How many ears of corn are there in total? __4__. Write the equation.

_____ × _____ = _____

Fill in the Answer

Fill in the blanks to complete the equations.

$2 \times \square = 6$ | $\square \times 2 = 6$ | $\square \times 2 = 6$

$3 \times 2 = \square$ | $2 \times 3 = \square$ | $3 \times 2 = \square$

$\square \times 3 = 6$ | $3 \times \square = 6$ | $\square \times 3 = 6$

$3 \times \square = 6$ | $2 \times 3 = \square$ | $2 \times \square = 6$

$\begin{array}{r} 2 \\ \times\ 3 \\ \hline \square \end{array}$ $\begin{array}{r} \square \\ \times\ 2 \\ \hline 6 \end{array}$ $\begin{array}{r} 3 \\ \times\ \square \\ \hline 6 \end{array}$ $\begin{array}{r} \square \\ \times\ 2 \\ \hline 6 \end{array}$ $\begin{array}{r} 2 \\ \times\ \square \\ \hline 6 \end{array}$ $\begin{array}{r} 3 \\ \times\ 2 \\ \hline \square \end{array}$

$\begin{array}{r} 3 \\ \times\ 2 \\ \hline \square \end{array}$ $\begin{array}{r} 2 \\ \times\ \square \\ \hline 6 \end{array}$ $\begin{array}{r} \square \\ \times\ 3 \\ \hline 6 \end{array}$ $\begin{array}{r} 2 \\ \times\ \square \\ \hline 6 \end{array}$ $\begin{array}{r} 3 \\ \times\ 2 \\ \hline \square \end{array}$ $\begin{array}{r} \square \\ \times\ 3 \\ \hline 6 \end{array}$

Activities

There are two fish.

Each fish blows three bubbles.

How many bubbles are there in all? _____

Use the numbers below to make equations.

2 3 6

2 × ◯ = ◯

3 × ◯ = ◯

◯ × ◯ = 6

◯ × 2 = ◯

Draw two bubbles for each fish.

_____ × _____ = _____

Write the equation to match the picture.

Solve the equations and color by number.

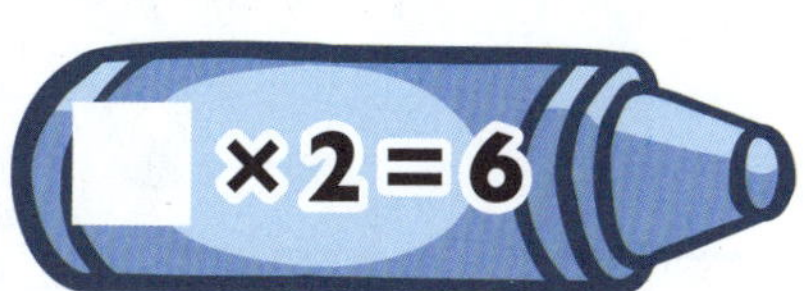

3	3	3	3	3	3	3	6	6	6	3	3	3
3	3	3	3	3	2	2	2	6	2	2	3	3
3	3	3	3	3	2	2	2	6	2	2	3	3
6	6	3	3	3	2	2	2	6	2	2	3	3
3	6	3	3	3	3	3	3	6	3	3	3	3
3	6	6	6	3	3	3	3	6	3	6	6	3
3	3	6	6	6	6	6	6	6	6	6	3	3
3	3	6	6	6	2	6	2	6	2	6	3	3
3	3	6	6	6	6	6	6	6	6	6	3	3
3	3	3	3	3	3	3	3	3	3	3	3	3

Fill in the Answer

Fill in the blanks to complete the equations.

___ × 4 = 8	4 × 2 = ___	___ × 4 = 8
2 × 4 = ___	4 × ___ = 8	4 × 2 = ___
4 × ___ = 8	___ × 2 = 8	___ × 2 = 8
2 × ___ = 8	2 × 4 = ___	2 × ___ = 8

2	___	2	___	2	2
× 4	× 2	× ___	× 4	× ___	× 4
___	8	8	8	8	___

4	2	___	2	4	___
× 2	× ___	× 4	× ___	× 2	× 4
___	8	8	8	___	8

Activities

There are four lights at the bottom of the mirror.

There are four lights on top of the mirror.

How many lights are there in all? _____

Use the numbers below to make equations.

2 4 8

___ × 4 = ___

4 × ___ = ___

___ × ___ = 8

___ × 2 = ___

Draw two buttons on each shirt.

______ × ______ = ______

Write the equation to match the picture.

How many bow ties are there in total? _____. Write the equation.

2 × 4 = 8

Fill in the Answer

Fill in the blanks to complete the equations.

2 × 5 = ___	2 × ___ = 10	5 × 2 = ___
5 × ___ = 10	___ × 5 = 10	2 × ___ = 10
___ × 2 = 10	5 × ___ = 10	___ × 5 = 10
5 × 2 = ___	___ × 2 = 10	2 × 5 = ___

2	___	2	___	2	2
× 5	× 2	× ___	× 5	× ___	× 5
___	10	10	10	10	___

5	2	___	2	5	___
× 2	× ___	× 5	× ___	× 2	× 5
___	10	10	10	___	10

Activities

There are two crowns.

There will be five gems on each crown.

How many gems are there in all? ____

Use the numbers below to make equations.

2 5 10

Draw two gems on each crown.

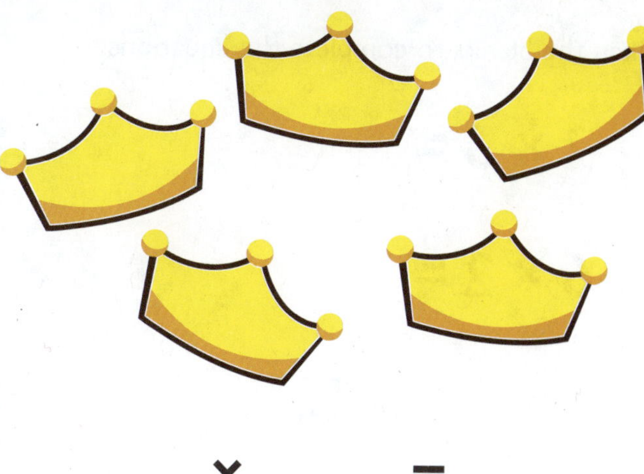

____ × ____ = ____

Write the equation to match the picture.

Solve the equations and color by number.

Fill in the Answer

Fill in the blanks to complete the equations.

2 × 6 = ____	____ × 6 = 12	____ × 2 = 12
6 × 2 = ____	6 × ____ = 12	6 × ____ = 12
2 × ____ = 12	____ × 6 = 12	6 × 2 = ____
____ × 2 = 12	2 × ____ = 12	2 × 6 = ____

2 × 6 = ____	____ × 2 = 12	6 × ____ = 12	____ × 2 = 12	2 × ____ = 12	2 × 6 = ____
6 × 2 = ____	2 × ____ = 12	____ × 6 = 12	2 × ____ = 12	6 × 2 = ____	____ × 6 = 12

Activities

There are six baskets. There are two carrots for each basket. How many carrots are there in all? ____

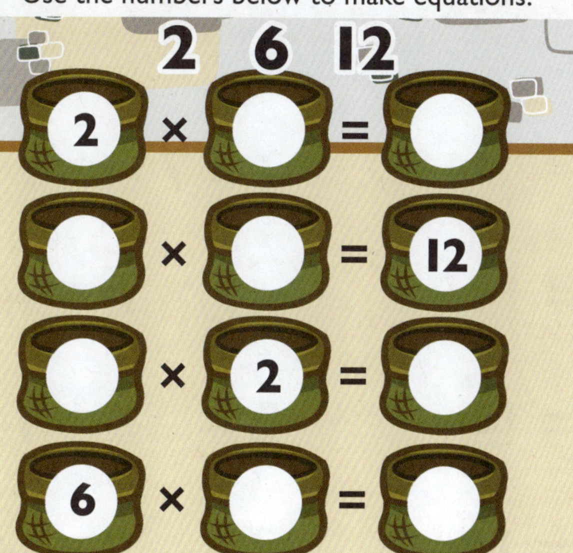

Draw two arrows in each quiver.

____ × ____ = ____

Write the equation to match the picture.

How many carrots are there in total? ____. Write the equation.

____ × ____ = ____

Fill in the Answer

Fill in the blanks to complete the equations.

☐ × 2 = 14	7 × 2 = ☐	2 × 7 = ☐
7 × 2 = ☐	☐ × 7 = 14	☐ × 7 = 14
2 × ☐ = 14	7 × ☐ = 14	2 × ☐ = 14
2 × 7 = ☐	☐ × 2 = 14	7 × ☐ = 14

☐ × 7 = 14	2 × 7 = ☐	2 × ☐ = 14	☐ × 7 = 14	7 × ☐ = 14	2 × 7 = ☐
7 × 2 = ☐	2 × ☐ = 14	☐ × 7 = 14	2 × ☐ = 14	7 × 2 = ☐	☐ × 7 = 14

Activities

There are two staffs.

There are seven notes per staff.

How many notes are there in all? ____

Draw two music notes on each staff.

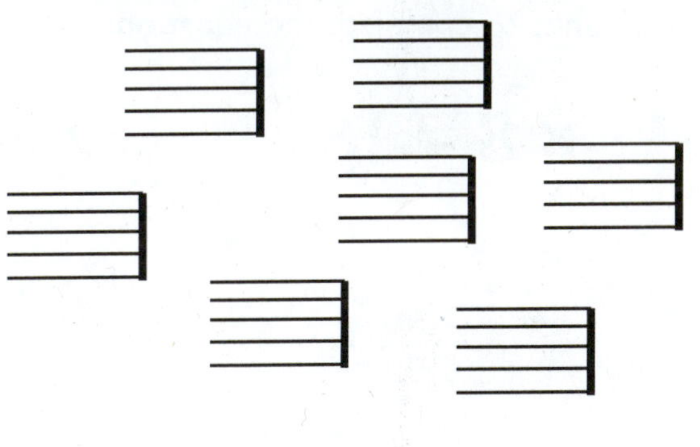

____ × ____ = ____

Write the equation to match the picture.

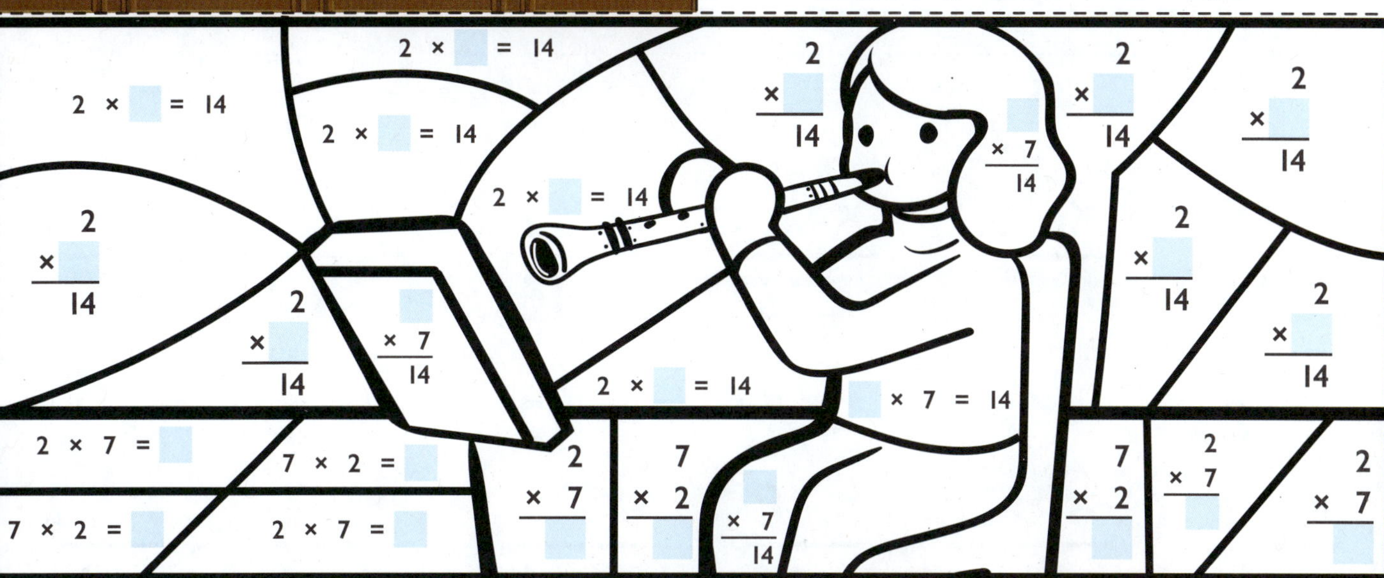

Solve the equations and color by number.

Fill in the Answer

Fill in the blanks to complete the equations.

___ × 8 = 16	2 × 8 = ___	___ × 2 = 16
8 × 2 = ___	2 × ___ = 16	2 × 8 = ___
2 × ___ = 16	___ × 2 = 16	___ × 8 = 16
8 × ___ = 16	8 × 2 = ___	8 × ___ = 16

___	2	8	___	2	2
× 8	× 8	× ___	× 2	× ___	× 8
16	___	16	16	16	___

8	8	___	2	8	___
× 2	× ___	× 8	× ___	× 2	× 2
___	16	16	16	___	16

Activities

There are eight hooks.

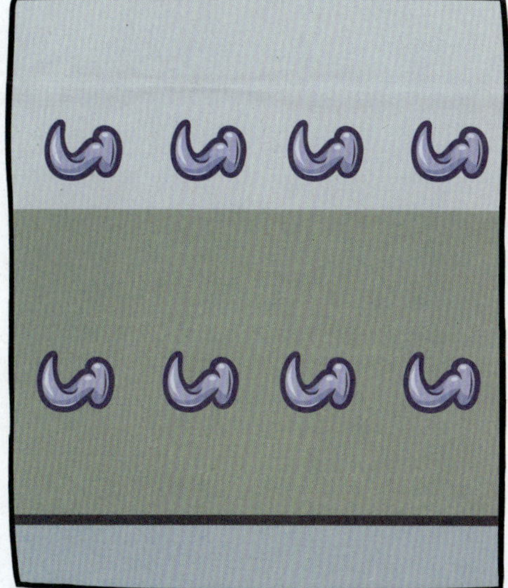

There are two sets of goggles on each hook.

How many sets of goggles are there in all? _____

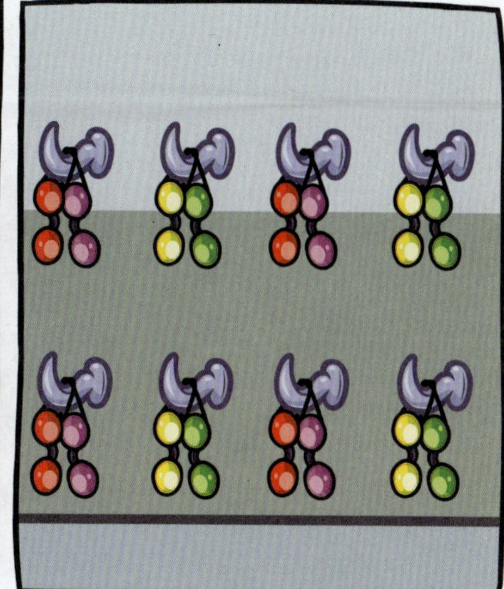

Draw eight eggs in each nest.

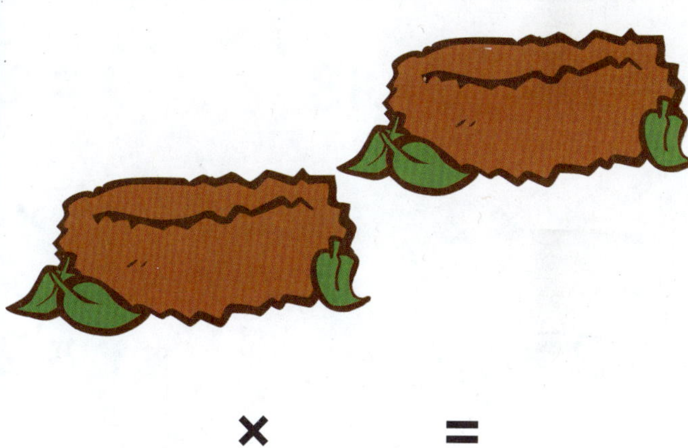

_____ × _____ = _____

Write the equation to match the picture.

How many hourglasses are there in total? _____. Write the equation.

_____ × _____ = _____

Fill in the Answer

Fill in the blanks to complete the equations.

$2 \times \square = 18$ | $9 \times 2 = \square$ | $2 \times \square = 18$

$\square \times 2 = 18$ | $\square \times 9 = 18$ | $2 \times 9 = \square$

$2 \times 9 = \square$ | $9 \times 2 = \square$ | $9 \times \square = 18$

$\square \times 9 = 18$ | $9 \times \square = 18$ | $\square \times 2 = 18$

$\begin{array}{r} 2 \\ \times\ 9 \\ \hline \square \end{array}$ $\begin{array}{r} \square \\ \times\ 2 \\ \hline 18 \end{array}$ $\begin{array}{r} 9 \\ \times\ \square \\ \hline 18 \end{array}$ $\begin{array}{r} \square \\ \times\ 2 \\ \hline 18 \end{array}$ $\begin{array}{r} 9 \\ \times\ \square \\ \hline 18 \end{array}$ $\begin{array}{r} 2 \\ \times\ 9 \\ \hline \square \end{array}$

$\begin{array}{r} 9 \\ \times\ 2 \\ \hline \square \end{array}$ $\begin{array}{r} 2 \\ \times\ \square \\ \hline 18 \end{array}$ $\begin{array}{r} \square \\ \times\ 9 \\ \hline 18 \end{array}$ $\begin{array}{r} 2 \\ \times\ \square \\ \hline 18 \end{array}$ $\begin{array}{r} 9 \\ \times\ 2 \\ \hline \square \end{array}$ $\begin{array}{r} \square \\ \times\ 9 \\ \hline 18 \end{array}$

Activities

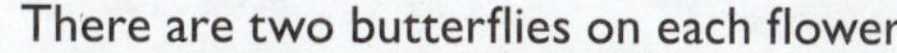

There are nine flowers.

There are two butterflies on each flower.

How many butterflies are there in all? ____

Use the numbers below to make equations.

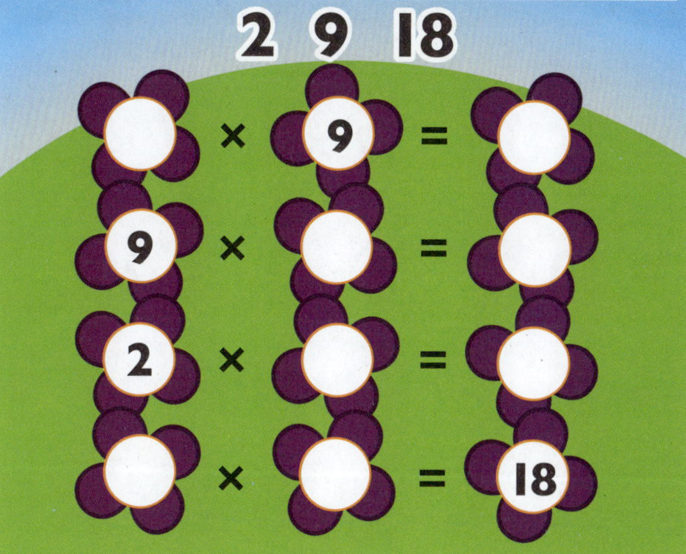

Draw nine apples in each basket.

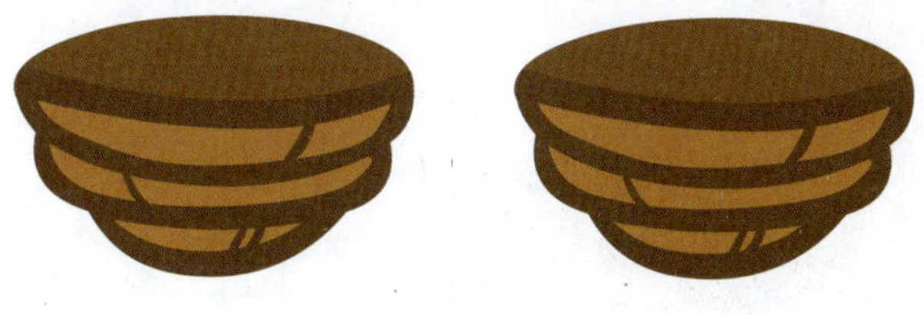

____ × ____ = ____

Write the equation to match the picture.

Solve the equations and color by number.

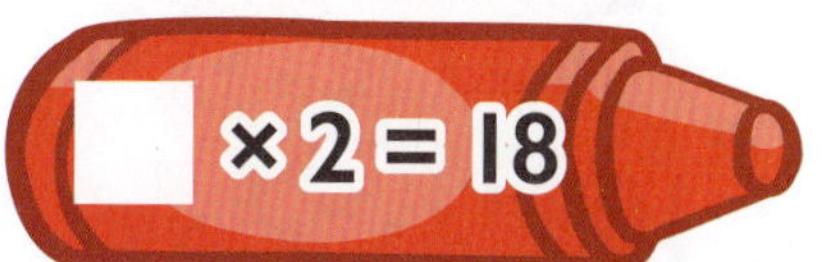

18	18	18	18	18	2	18	2	18	18	18	18	18
18	18	9	9	9	2	18	2	9	9	9	18	18
18	9	9	9	9	9	2	9	9	9	9	9	18
18	9	2	2	2	9	2	9	2	2	2	9	18
18	9	9	2	9	9	2	9	9	2	9	9	18
18	18	9	9	9	9	2	9	9	9	9	18	18
18	18	18	18	18	9	2	9	18	18	18	18	18
18	18	18	18	9	9	2	9	9	18	18	18	18
18	18	18	9	9	9	2	9	9	9	18	18	18
18	18	18	18	9	9	18	9	9	18	18	18	18

Fill in the Answer

Fill in the blanks to complete the equations.

2 × ___ = 20	2 × 10 = ___	2 × 10 = ___
___ × 10 = 20	2 × ___ = 20	10 × ___ = 20
10 × ___ = 20	___ × 2 = 20	___ × 10 = 20
___ × 2 = 20	10 × 2 = ___	10 × 2 = ___

2	___	10	___	2	2
× 10	× 2	× ___	× 2	× ___	× 10
___	20	20	20	20	___

10	2	___	2	10	___
× 2	× ___	× 10	× ___	× 2	× 2
___	20	20	20	___	20

Activities

There are two cacti. Each cactus has ten thorns. How many thorns are there in all? ______

Use the numbers below to make equations.

2 10 20

	×	10	=	
10	×		=	
	×		=	20
	×	2	=	

Draw two desert flowers on each cactus.

______ × ______ = ______

Write the equation to match the picture.

Solve the equations and color by number.

20

Fill in the Answer

Fill in the blanks to complete the equations.

2 × 11 = ___	11 × ___ = 22	2 × ___ = 22
2 × ___ = 22	___ × 2 = 22	11 × 2 = ___
___ × 11 = 22	2 × 11 = ___	___ × 2 = 22
11 × 2 = ___	___ × 11 = 22	11 × ___ = 22

2	11	2	___	2	___
× ___	× 2	× 11	× 2	× ___	× 11
22	___	___	22	22	22

2	11	___	2	11	___
× 11	× ___	× 11	× ___	× 2	× 2
___	22	22	22	___	22

Activities

There are two trees.

Each tree has eleven bells.

How many bells are there in all? _____

Use the numbers below to make equations.

Draw two leaves on each flower.

_____ × _____ = _____

Write the equation to match the picture.

How many bells are there in total? _____. Write the equation.

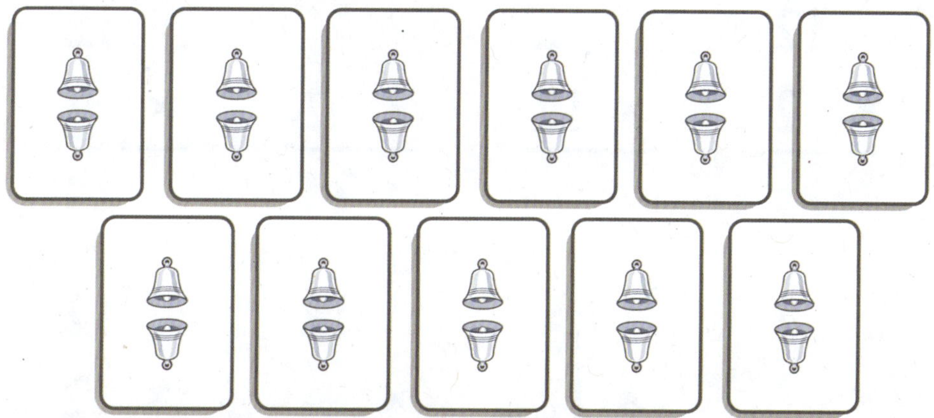

_____ × _____ = _____

Fill in the Answer

Fill in the blanks to complete the equations.

2 × ___ = 24	___ × 12 = 24	2 × 12 = ___
2 × 12 = ___	2 × ___ = 24	12 × ___ = 24
12 × ___ = 24	___ × 2 = 24	___ × 12 = 24
___ × 2 = 24	12 × 2 = ___	12 × 2 = ___

2	12	2	___	12	___
× ___	× 2	× 12	× 2	× ___	× 12
24	___	___	24	24	24

2	2	___	12	12	___
× 12	× ___	× 12	× ___	× 2	× 2
___	24	24	24	___	24

Activities

There are two bowls.

Each bowl has twelve cubes of sugar.

How many cubes are there in all? ____

Use the numbers below to make equations.

2 12 24

☐ × 12 = ☐

12 × ☐ = ☐

☐ × ☐ = 24

☐ × 2 = ☐

Draw two sugar cubes on each spoon.

____ × ____ = ____

Write the equation to match the picture.

☐ × 12 = 24

12 × ☐ = 24

12 × 2 = ☐

☐ × 12 = 24

☐ × 2 = 24

2 × ☐ = 24

12 × ☐ = 24

12 × 2 = ☐

Solve the equations and color by number.

2 12 24

Fill in the Answer

Fill in the blanks to complete the equations.

____ × 3 = 9	3 × ____ = 9	3 × 3 = ____
3 × ____ = 9	____ × 3 = 9	3 × ____ = 9
3 × 3 = ____	3 × 3 = ____	____ × 3 = 9
____ × 3 = 9	3 × ____ = 9	3 × 3 = ____

---

3 × ____ 9	3 × 3 ____	____ × 3 9	3 × 3 ____	3 × ____ 9	____ × 3 9

---

3 × 3 ____	3 × ____ 9	____ × 3 9	3 × ____ 9	3 × 3 ____	____ × 3 9

Activities

There are three mounds.

Each mound has three clams.

How many clams are there in all? _____

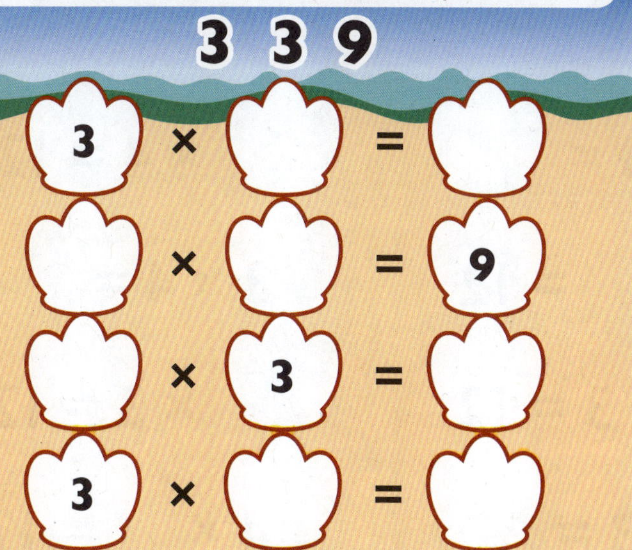

Draw three pearls in each clam.

_____ × _____ = _____

Write the equation to match the picture.

Solve the equations and color by number.

9	9	3	3	3	3	3	3	3	3	3	9	9
9	3	3	3	3	3	3	3	3	3	3	3	9
9	3	9	3	3	3	3	3	3	3	9	3	9
9	9	9	3	3		3		3	3	9	9	9
3	9	3	3	3	9	3	9	3	3	3	9	3
3	9	3	3	9	9	9	9	9	3	3	9	3
3	9	9	9	9	9	9	9	9	9	9	9	3
3	3	3	3	9	9	9	9	9	3	3	3	3
3	3	9	9	9	9	9	9	9	9	9	3	3
3	3	9	3	3	3	3	3	3	3	9	3	3

Fill in the Answer

Fill in the blanks to complete the equations.

3 × ___ = 12	4 × 3 = ___	___ × 4 = 12
___ × 4 = 12	3 × ___ = 12	3 × 4 = ___
3 × 4 = ___	___ × 3 = 12	___ × 3 = 12
4 × ___ = 12	4 × 3 = ___	4 × ___ = 12

3	___	4	___	3	3
× 4	× 3	× ___	× 4	× ___	× 4
___	12	12	12	12	___

4	3	___	4	4	___
× 3	× ___	× 4	× ___	× 3	× 3
___	12	12	12	___	12

Activities

There are four shelves.

Each shelf has three brushes.

How many brushes are there in all? ____

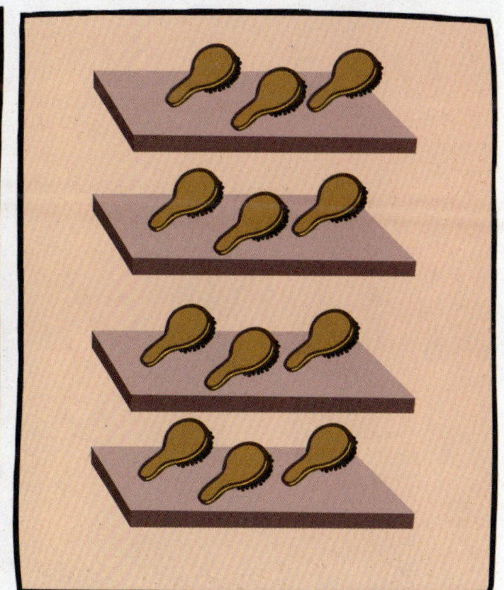

Use the numbers below to make equations.

3 4 12

☐	×	4	=	☐
4	×	☐	=	☐
☐	×	☐	=	12
☐	×	3	=	☐

Draw three hair combs in each cup.

_____ × _____ = _____

Write the equation to match the picture.

How many blow dryers are there in total? _____. Write the equation.

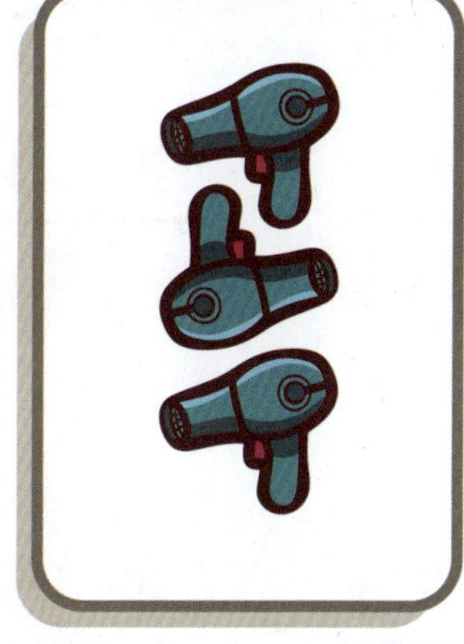

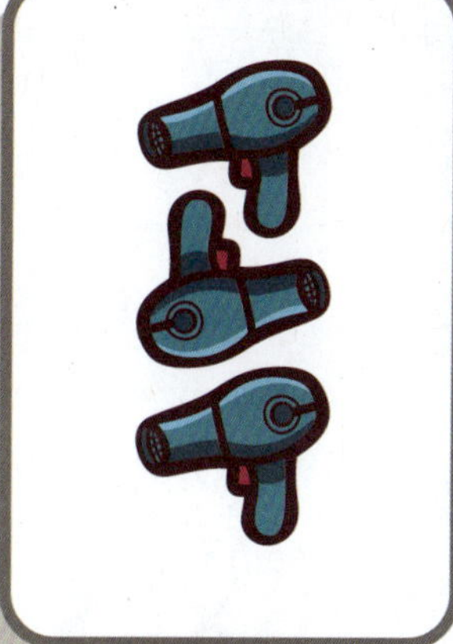

_____ × _____ = _____

Fill in the Answer

Fill in the blanks to complete the equations.

3 × 5 = ___	___ × 5 = 15	5 × ___ = 15
5 × ___ = 15	5 × 3 = ___	3 × 5 = ___
___ × 3 = 15	3 × ___ = 15	___ × 5 = 15
3 × ___ = 15	___ × 3 = 15	5 × 3 = ___

3	___	3	___	5	5
× 5	× 5	× ___	× 5	× ___	× 3
___	15	15	15	15	___

5	3	___	5	3	___
× 3	× ___	× 5	× ___	× 5	× 5
___	15	15	15	___	15

Activities

There are three match boxes. Each box has five matches. How many matches are there in all? ____

Draw three gems in each cart.

____ × ____ = ____

Write the equation to match the picture.

Solve the equations and color by number.

15

Fill in the Answer

Fill in the blanks to complete the equations.

3 × ___ = 18	6 × 3 = ___	6 × 3 = ___
3 × 6 = ___	___ × 6 = 18	___ × 6 = 18
___ × 3 = 18	3 × ___ = 18	6 × ___ = 18
6 × ___ = 18	3 × 6 = ___	___ × 3 = 18

3	___	6	___	3	3
× 6	× 6	× ___	× 6	× ___	× 6
___	18	18	18	18	___

6	3	___	6	3	___
× 3	× ___	× 6	× ___	× 6	× 3
___	18	18	18	___	18

Activities

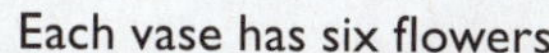

There are three vases.

Each vase has six flowers.

How many flowers are there in all? ______

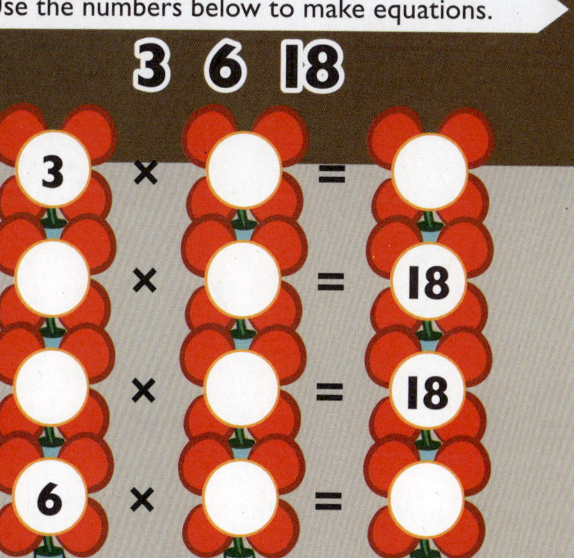

Draw three flowers in each pot.

______ × ______ = ______

Write the equation to match the picture.

Solve the equations and color by number.

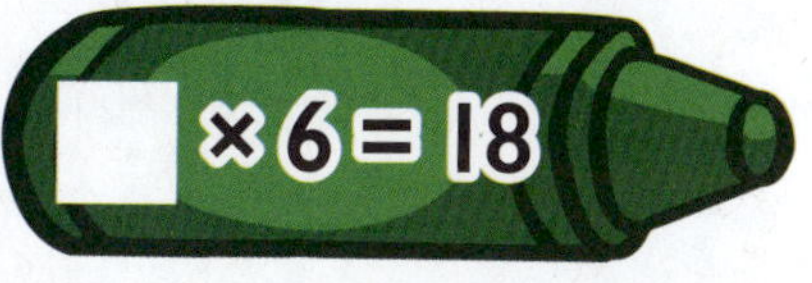

☐ × 6 = 18

6	6	6	6	6	18	18	18	6	6	6	6	6
6	6	18	18	18	18	18	18	18	18	18	6	6
6	18	18	18	18	6	6	6	18	18	18	18	6
6	18	18	18	6	6	6	6	6	18	18	18	6
6	18	18	18	6	6	6	6	6	18	18	18	6
6	18	18	18	18	6	6	6	18	18	18	18	6
6	6	18	18	18	18	18	18	18	18	18	6	6
6	6	6	6	6	18	18	18	6	6	6	6	6
6	3	6	3	6	3	3	3	6	3	6	3	6
6	3	6	3	6	3	3	3	6	3	6	3	6

Fill in the Answer

Fill in the blanks to complete the equations.

$3 \times 7 = \square$ $7 \times 3 = \square$ $\square \times 7 = 21$

$7 \times \square = 21$ $\square \times 7 = 21$ $7 \times 3 = \square$

$\square \times 3 = 21$ $3 \times 7 = \square$ $\square \times 3 = 21$

$3 \times \square = 21$ $3 \times \square = 21$ $7 \times \square = 21$

$7 \times 3 = \square$ $\square \times 7 = 21$ $3 \times \square = 21$ $\square \times 3 = 21$ $7 \times \square = 21$ $3 \times 7 = \square$

$3 \times \square = 21$ $7 \times 3 = \square$ $\square \times 7 = 21$ $3 \times \square = 21$ $7 \times 3 = \square$ $\square \times 7 = 21$

Activities

There are three bundles.

Each bundle has seven sticks.

How many sticks are there in all? _____

Use the numbers below to make equations.

Draw three snowballs in each snow cone.

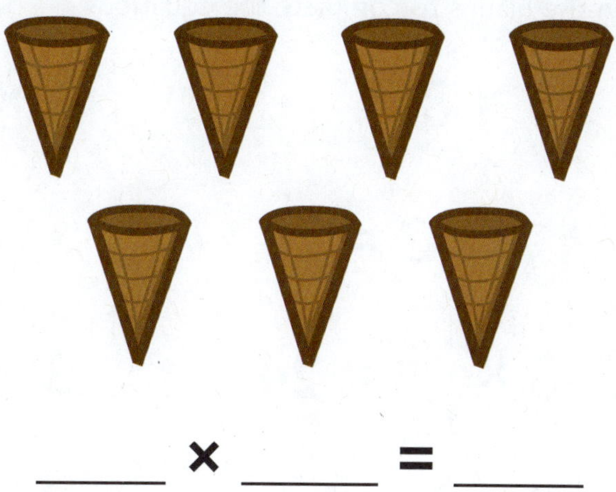

_____ × _____ = _____

Write the equation to match the picture.

How many goggles are there in total? _____. Write the equation.

_____ × _____ = _____

Fill in the Answer

Fill in the blanks to complete the equations.

3 × 8 = ___	3 × ___ = 24	___ × 8 = 24
___ × 3 = 24	___ × 8 = 24	8 × ___ = 24
8 × ___ = 24	3 × 8 = ___	8 × 3 = ___
8 × 3 = ___	8 × ___ = 24	___ × 8 = 24

8	3	___	8	3	___
× ___	× 8	× 8	× 3	× ___	× 3
24	___	24	___	24	24

3	8	___	3	8	___
× 8	× ___	× 3	× ___	× 3	× 8
___	24	24	24	___	24

Activities

There are three shovels. Each shovel scoops eight snowballs. How many balls are there in all? ______

Use the numbers below to make equations.

3 8 24

___ × 8 = ___

8 × ___ = ___

___ × ___ = 24

___ × 3 = ___

Draw three buttons on each snowman.

______ × ______ = ______

Write the equation to match the picture.

Solve the equations and color by number.

Fill in the Answer

Fill in the blanks to complete the equations.

9 × ___ = 27   ___ × 9 = 27   ___ × 3 = 27

9 × 3 = ___   9 × 3 = ___   3 × 9 = ___

___ × 3 = 27   3 × ___ = 27   ___ × 9 = 27

3 × ___ = 27   3 × 9 = ___   9 × ___ = 27

3	3	___	3	3	___
× ___	× 9	× 3	× 9	× ___	× 3
27	___	27	___	27	27

9	9	___	3	3	___
× 3	× ___	× 3	× ___	× 9	× 3
___	27	27	27	___	27

Activities

There are three balls in a tube.

There are nine tubes.

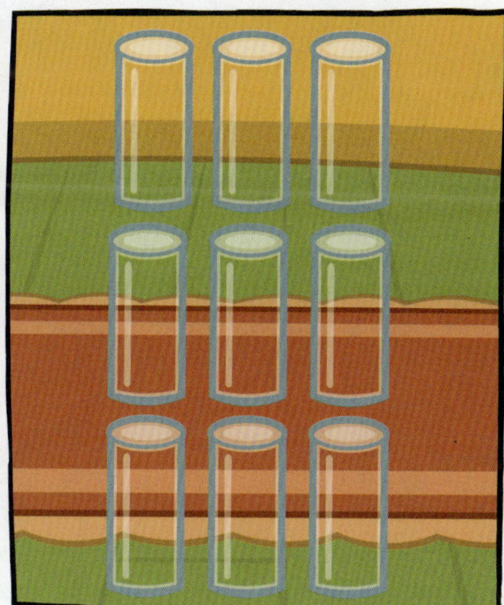

How many balls are there in all? _____

Use the numbers below to make equations.

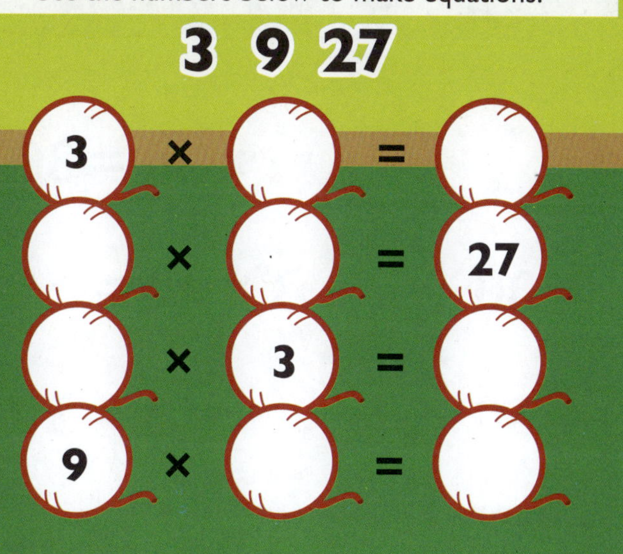

Draw three balls of yarn in each basket.

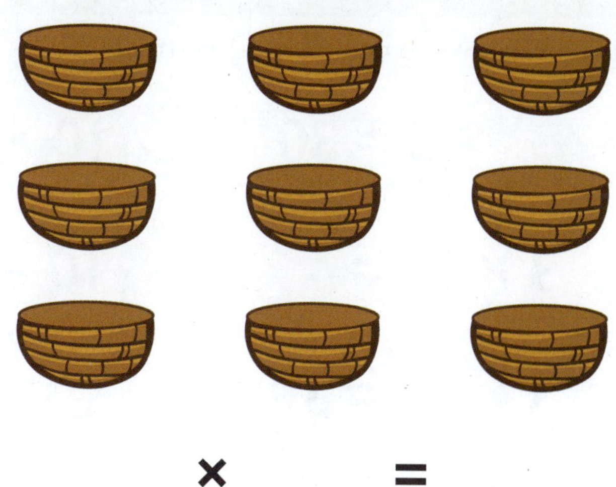

_____ × _____ = _____

Write the equation to match the picture.

Solve the equations and color by number.

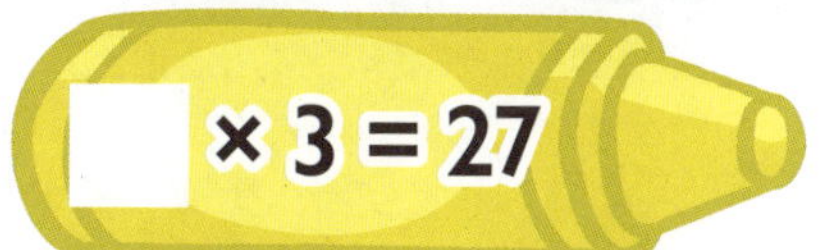

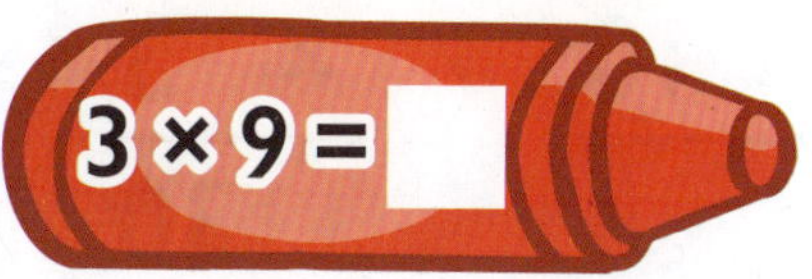

9	9	9	27	9	9	9	9	9	27	9	9	9
9	9	9	27	27	9	9	9	27	27	9	9	9
9	9	9	27	27	27	27	27	27	27	9	9	9
9	9	9	27	3	27	27	27	3	27	9	9	9
9	9	9	27	3	27	27	27	3	27	9	9	9
9	9	9	27	3	27	27	27	3	27	9	9	9
9	9	9	27	27	27	27	27	27	27	9	9	9
9	3	3	3	27	3	3	3	27	3	3	3	9
9	9	9	9	27	27	3	27	27	9	9	9	9
9	9	9	9	9	27	27	27	9	9	9	9	9

Fill in the Answer

Fill in the blanks to complete the equations.

$\square \times 3 = 30$	$10 \times 3 = \square$	$\square \times 10 = 30$
$3 \times 10 = \square$	$10 \times \square = 30$	$3 \times 10 = \square$
$3 \times \square = 30$	$\square \times 10 = 30$	$\square \times 3 = 30$
$10 \times \square = 30$	$10 \times 3 = \square$	$3 \times \square = 30$

3	10	$\square$	3	3	$\square$
× $\square$	× 3	× 3	× 10	× $\square$	× 3
30	$\square$	30	$\square$	30	30

3	3	$\square$	10	3	$\square$
× 10	× $\square$	× 10	× $\square$	× 10	× 3
$\square$	30	30	30	$\square$	30

Activities

There are three bags.

Each bag has ten apples.

How many apples are there in all? _____

Use the numbers below to make equations.

3 10 30

_____ × 10 = _____

_____ × _____ = 30

3 × _____ = _____

10 × _____ = _____

Draw three string beans in each can.

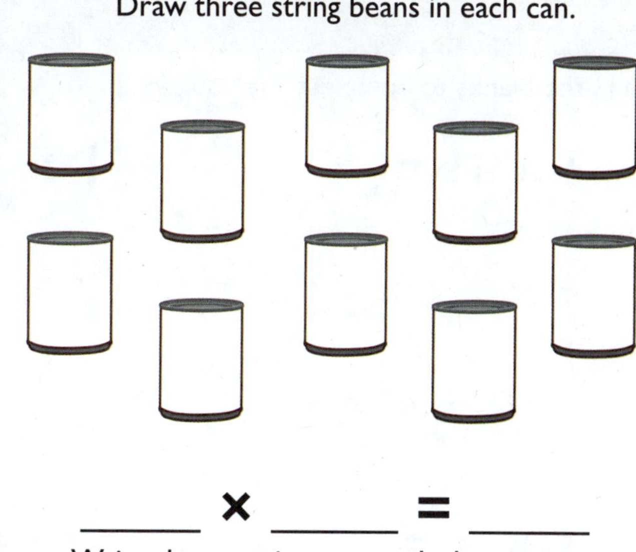

_____ × _____ = _____

Write the equation to match the picture.

How many apples are there in total? _____. Write the equation.

_____ × _____ = _____

Fill in the Answer

Fill in the blanks to complete the equations.

3 × 11 = ___	11 × ___ = 33	3 × 11 = ___
3 × ___ = 33	___ × 11 = 33	11 × ___ = 33
___ × 3 = 33	3 × ___ = 33	___ × 11 = 33
11 × 3 = ___	___ × 3 = 33	11 × 3 = ___

3	3	___	11	3	___
× ___	× 11	× 3	× 3	× ___	× 11
33	___	33	___	33	33

11	3	___	3	3	___
× 3	× ___	× 11	× ___	× 11	× 3
___	33	33	33	___	33

Activities

There are three jellyfish.

Each jellyfish has eleven tentacles.

How many tentacles are there in all? ____

Draw three bubbles on each coral.

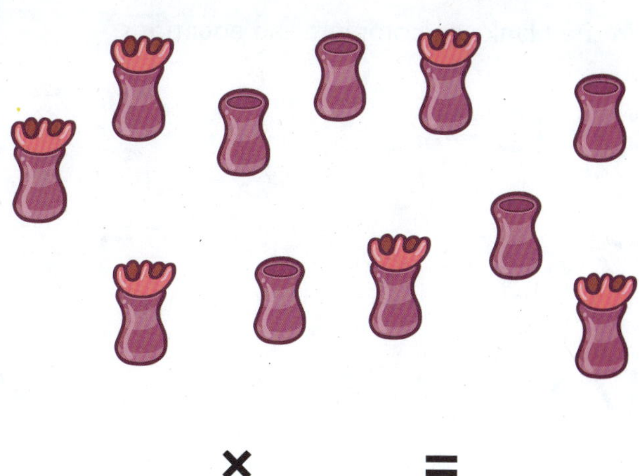

____ × ____ = ____

Write the equation to match the picture.

Solve the equations and color by number.

Fill in the Answer

Fill in the blanks to complete the equations.

3 × 12 = ___	___ × 12 = 36	___ × 12 = 36
12 × 3 = ___	3 × ___ = 36	3 × ___ = 36
12 × ___ = 36	___ × 3 = 36	12 × 3 = ___
___ × 3 = 36	12 × ___ = 36	3 × 12 = ___

3	___	3	___	12	3
× 12	× 12	× ___	× 3	× ___	× 12
___	36	36	36	36	___

3	12	___	3	12	___
× ___	× 3	× 12	× ___	× 3	× 12
36	___	36	36	___	36

Activities

There are three harps.

Each harp has twelve strings.

How many strings are there in all? _____

Use the numbers below to make equations.

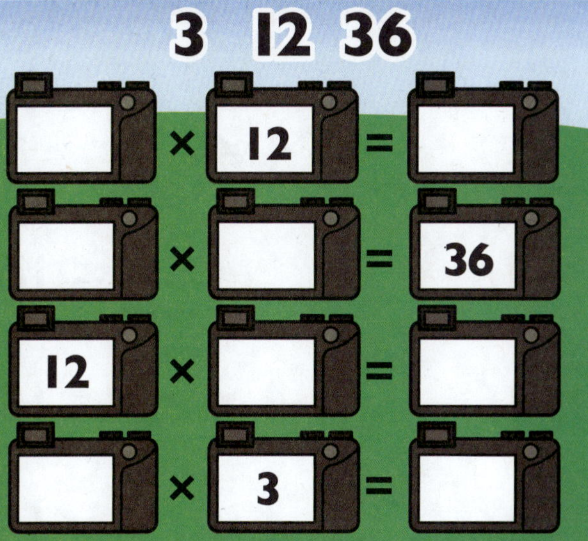

Fill in the missing numbers and color by number.

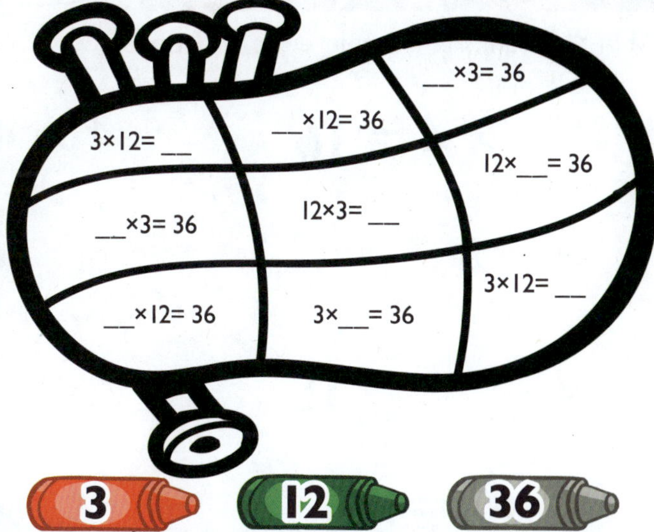

Draw three pipes on each bag.

Write the equation to match the picture.

_____ × _____ = _____

Fill in the Answer

Fill in the blanks to complete the equations.

☐ × 4 = 16	4 × 4 = ☐	4 × 4 = ☐
4 × 4 = ☐	☐ × 4 = 16	☐ × 4 = 16
4 × ☐ = 16	4 × ☐ = 16	4 × ☐ = 16
4 × 4 = ☐	☐ × 4 = 16	4 × ☐ = 16

4	☐	4	☐	4	4
× 4	× 4	× ☐	× 4	× ☐	× 4
☐	16	16	16	16	☐

4	4	☐	4	4	☐
× ☐	× 4	× 4	× ☐	× 4	× 4
16	☐	16	16	☐	16

Activities

There are four arcade games.

Each game has four buttons.

How many buttons are there in all? ____

Draw four coins on each table top.

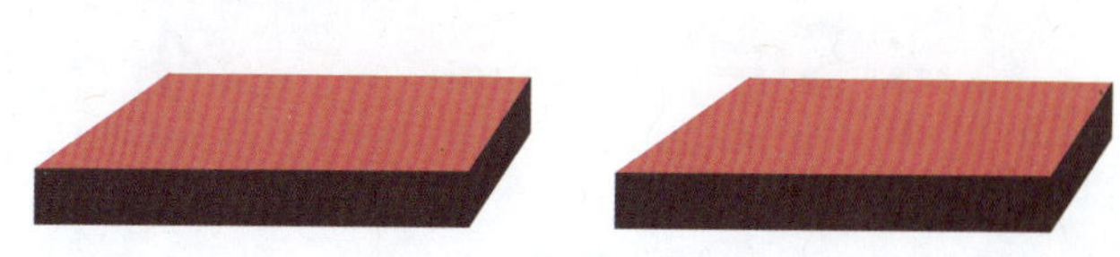

____ × ____ = ____

Write the equation to match the picture.

How many helmets are there in total? ____. Write the equation.

____ × ____ = ____

Fill in the Answer

Fill in the blanks to complete the equations.

☐ × 4 = 20	4 × 5 = ☐	☐ × 4 = 20
4 × 5 = ☐	4 × ☐ = 20	5 × 4 = ☐
5 × ☐ = 20	☐ × 5 = 20	☐ × 5 = 20
4 × ☐ = 20	5 × 4 = ☐	5 × ☐ = 20

$$\begin{array}{r}4\\ \times\ \square\\ \hline 20\end{array}\quad \begin{array}{r}4\\ \times\ 5\\ \hline \square\end{array}\quad \begin{array}{r}\square\\ \times\ 5\\ \hline 20\end{array}\quad \begin{array}{r}5\\ \times\ 4\\ \hline \square\end{array}\quad \begin{array}{r}4\\ \times\ \square\\ \hline 20\end{array}\quad \begin{array}{r}\square\\ \times\ 5\\ \hline 20\end{array}$$

$$\begin{array}{r}5\\ \times\ 4\\ \hline \square\end{array}\quad \begin{array}{r}4\\ \times\ \square\\ \hline 20\end{array}\quad \begin{array}{r}\square\\ \times\ 5\\ \hline 20\end{array}\quad \begin{array}{r}5\\ \times\ \square\\ \hline 20\end{array}\quad \begin{array}{r}4\\ \times\ 5\\ \hline \square\end{array}\quad \begin{array}{r}\square\\ \times\ 4\\ \hline 20\end{array}$$

Activities

There are four heads of lettuce.

There are five baskets.

How many heads of lettuce are there in all? ______

Use the numbers below to make equations.

4 5 20

$\square \times 5 = \square$

$\square \times \square = 20$

$\square \times 4 = \square$

$5 \times \square = \square$

Draw five tomatoes on each spear.

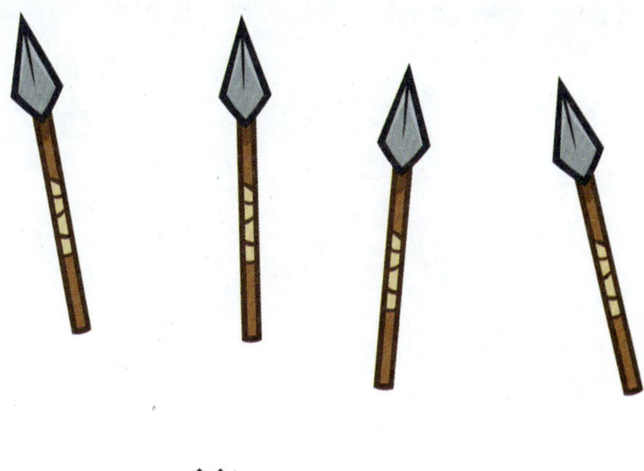

______ × ______ = ______

Write the equation to match the picture.

Solve the equations and color by number.

Fill in the Answer

Fill in the blanks to complete the equations.

6 × ___ = 24	6 × 4 = ___	6 × ___ = 24
___ × 6 = 24	___ × 4 = 24	4 × 6 = ___
4 × 6 = ___	6 × 4 = ___	4 × ___ = 24
___ × 4 = 24	4 × ___ = 24	___ × 6 = 24

4	4	___	6	4	___
× ___	× 6	× 4	× 4	× ___	× 4
24	___	24	___	24	24

6	6	___	4	6	___
× 4	× ___	× 4	× ___	× 4	× 4
___	24	24	24	___	24

Activities

He has four loafs of bread.

He breaks each loaf into pieces of six.

How many pieces are there in all? ______

Use the numbers below to make equations.

Draw four bread crumbs for each bird.

______ × ______ = ______

Write the equation to match the picture.

Solve the equations and color by number.

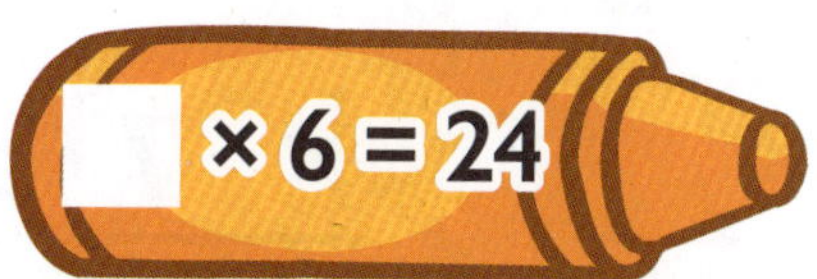

6	6	6	6	6	6	24	24	24	6	6	6	6
6	6	6	6	6	24	24	24	24	24	6	6	6
6	6	6	6	6		24		24	24	6	6	6
6	4	6	6	6		24		24	24	6	6	6
6	4	4	4	4	4	4	24	24	24	6	6	6
6	4	4	4	4	4	4	4	24	24	6	6	6
6	6	4	4	4	4	24	24	24	24	6	6	6
6	6	6	6	6	6	6	24	24	24	6	6	6
6	6	6	6	6	6	6	24	24	24	6	6	6
6	6	6	6	6	6	6	24	24	24	6	6	6

Fill in the Answer

Fill in the blanks to complete the equations.

4 × ___ = 28	7 × 4 = ___	7 × 4 = ___
___ × 4 = 28	7 × ___ = 28	4 × ___ = 28
7 × ___ = 28	___ × 7 = 28	___ × 4 = 28
___ × 7 = 28	4 × 7 = ___	4 × 7 = ___

4	4	___	7	4	___
× ___	× 7	× 7	× 4	× ___	× 7
28	___	28	___	28	28

7	7	___	4	4	___
× 4	× ___	× 4	× ___	× 7	× 7
___	28	28	28	___	28

Activities

There are four ducks.

Each duck gets seven crumbs.

How many crumbs are there in all? ____

Draw four seeds in each bag.

____ × ____ = ____

Write the equation to match the picture.

How many ducks are there in total? ____. Write the equation.

____ × ____ = ____

Fill in the Answer

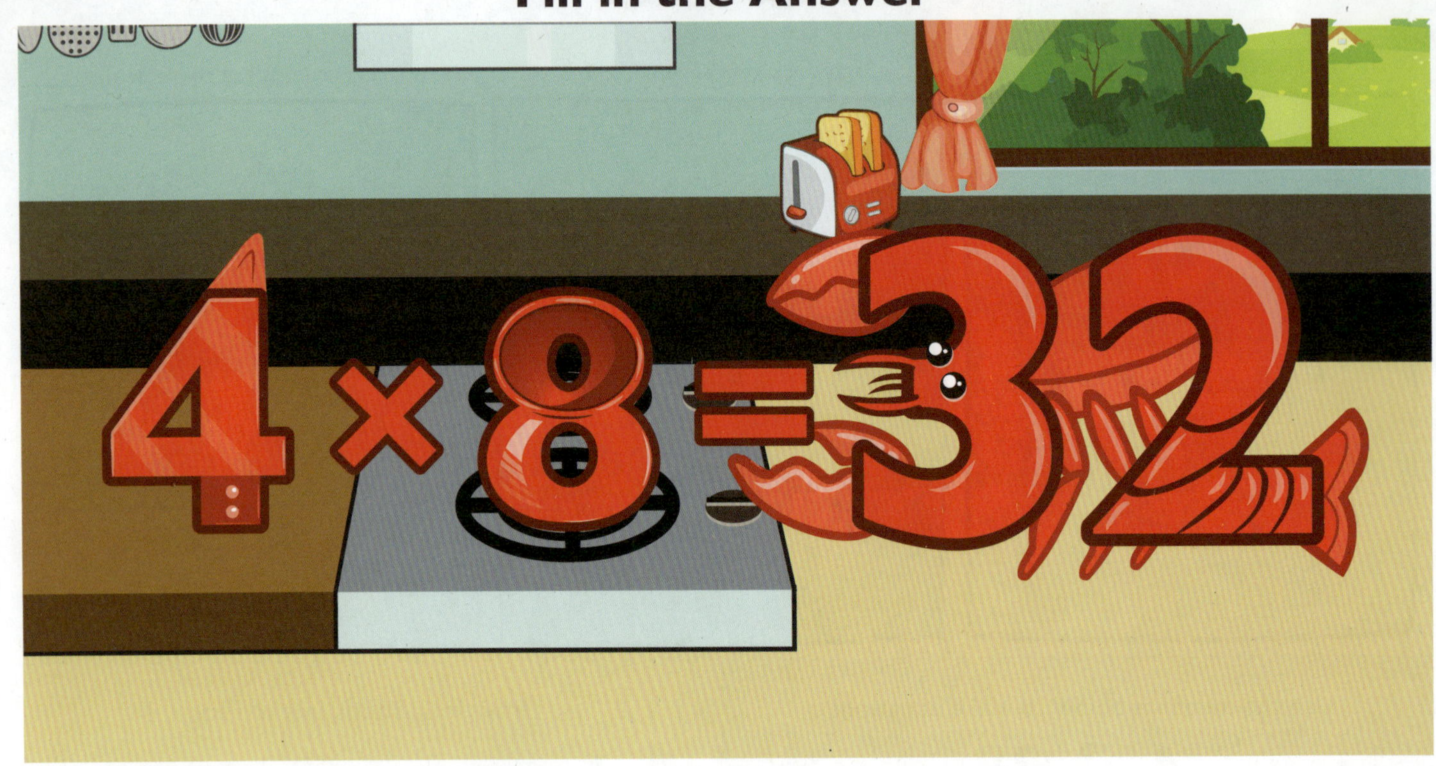

Fill in the blanks to complete the equations.

$8 \times 4 = \square$	$4 \times \square = 32$	$8 \times \square = 32$
$8 \times \square = 32$	$\square \times 4 = 32$	$4 \times 8 = \square$
$\square \times 8 = 32$	$4 \times 8 = \square$	$\square \times 4 = 32$
$8 \times 4 = \square$	$\square \times 8 = 32$	$4 \times \square = 32$

$$\begin{array}{r} 8 \\ \times\ \square \\ \hline 32 \end{array} \qquad \begin{array}{r} 4 \\ \times\ 8 \\ \hline \square \end{array} \qquad \begin{array}{r} \square \\ \times\ 8 \\ \hline 32 \end{array} \qquad \begin{array}{r} 8 \\ \times\ 4 \\ \hline \square \end{array} \qquad \begin{array}{r} 4 \\ \times\ \square \\ \hline 32 \end{array} \qquad \begin{array}{r} \square \\ \times\ 8 \\ \hline 32 \end{array}$$

$$\begin{array}{r} 8 \\ \times\ 4 \\ \hline \square \end{array} \qquad \begin{array}{r} 4 \\ \times\ \square \\ \hline 32 \end{array} \qquad \begin{array}{r} \square \\ \times\ 8 \\ \hline 32 \end{array} \qquad \begin{array}{r} 4 \\ \times\ \square \\ \hline 32 \end{array} \qquad \begin{array}{r} 8 \\ \times\ 4 \\ \hline \square \end{array} \qquad \begin{array}{r} \square \\ \times\ 4 \\ \hline 32 \end{array}$$

Activities

There are eight bowls. Each bowl has four pieces of carrot. How many pieces are there in all? ______

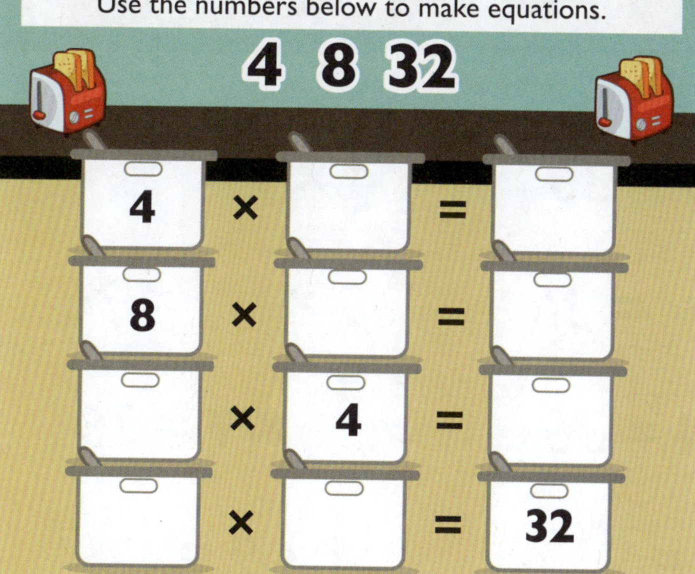

There are 8 buttons. Each button has 4 holes.

How many holes are there in all? _____

Draw eight zucchini in each pot.

Write the equation to match the picture. ______ × ______ = ______

Fill in the Answer

Fill in the blanks to complete the equations.

$4 \times \square = 36$ $\square \times 9 = 36$ $4 \times 9 = \square$

$9 \times 4 = \square$ $9 \times \square = 36$ $9 \times \square = 36$

$4 \times \square = 36$ $\square \times 4 = 36$ $\square \times 9 = 36$

$\square \times 4 = 36$ $9 \times 4 = \square$ $4 \times 9 = \square$

$4 \times \square = 36$ $4 \times 9 = \square$ $\square \times 9 = 36$ $9 \times 4 = \square$ $4 \times \square = 36$ $\square \times 9 = 36$

$9 \times 4 = \square$ $4 \times \square = 36$ $\square \times 9 = 36$ $4 \times \square = 36$ $4 \times 9 = \square$ $\square \times 4 = 36$

Activities

There are four jellyfish.

Each jellyfish get nine coins.

How many coins are there in all? ______

Use the numbers below to make equations.

4 × ◯ = ◯

9 × ◯ = ◯

◯ × ◯ = 36

◯ × 4 = ◯

There are 4 holes per die and 9 dice.

How many holes are there? ______

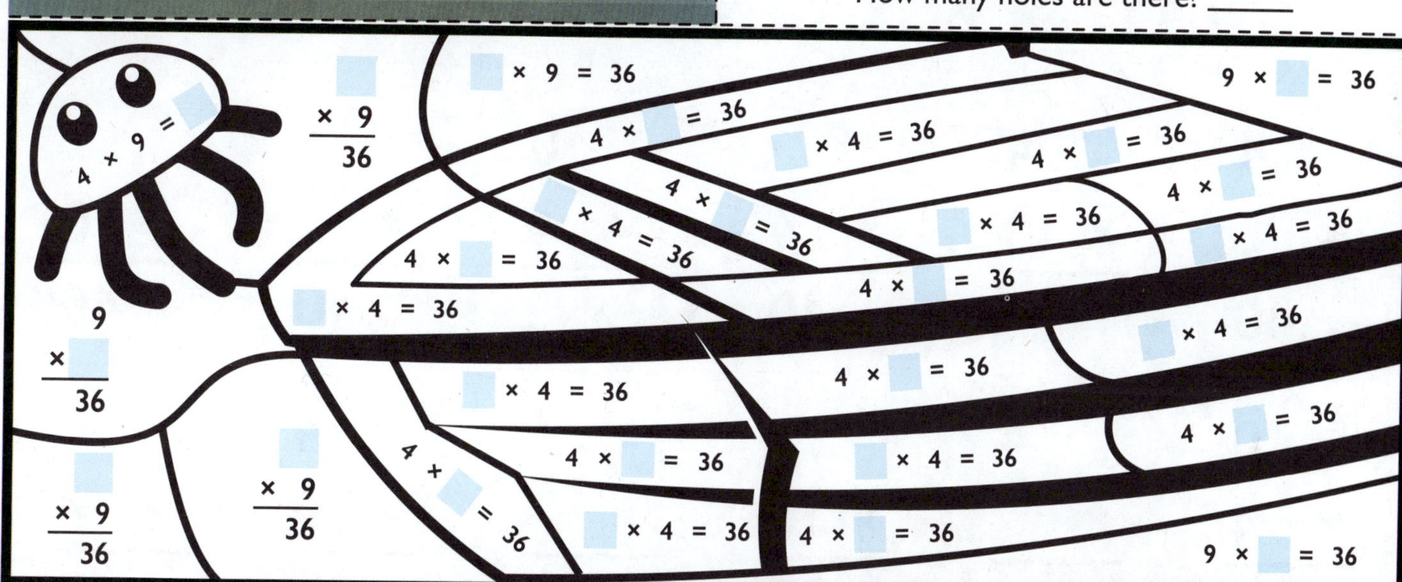

Solve the equations and color by number.

Fill in the Answer

Fill in the blanks to complete the equations.

$\square \times 4 = 40$	$4 \times \square = 40$	$4 \times 10 = \square$
$10 \times \square = 40$	$\square \times 4 = 40$	$4 \times \square = 40$
$4 \times 10 = \square$	$10 \times 4 = \square$	$\square \times 10 = 40$
$\square \times 10 = 40$	$10 \times \square = 40$	$10 \times 4 = \square$

4	4	$\square$	10	4	$\square$
× $\square$	× 10	× 4	× 4	× $\square$	× 10
40	$\square$	40	$\square$	40	40

10	4	$\square$	4	4	$\square$
× 4	× $\square$	× 10	× $\square$	× 10	× 4
$\square$	40	40	40	$\square$	40

Activities

There are four balloons.

Each balloon gets ten ribbons.

How many ribbons are there in all? ____

Use the numbers below to make equations.

4 10 40

4 × ___ = ___

___ × 4 = ___

10 × ___ = ___

___ × ___ = 40

Circle all of the 4, 10, 40 fact families.

4	10	40	8	9
12	10	3	13	11
10	4	40	25	40
18	4	55	4	45
10	4	10	40	72

Draw four balloons for each child.

Write the equation to match the picture. ____ × ____ = ____

Fill in the Answer

Fill in the blanks to complete the equations.

4 × ___ = 44   4 × 11 = ___   ___ × 11 = 44

___ × 11 = 44   4 × ___ = 44   4 × 11 = ___

11 × 4 = ___   ___ × 4 = 44   ___ × 4 = 44

11 × ___ = 44   11 × 4 = ___   11 × ___ = 44

4	___	4	___	4	11
× 11	× 11	× ___	× 4	× ___	× 4
___	44	44	44	44	___

4	4	___	4	11	___
× ___	× 11	× 11	× ___	× 4	× 4
44	___	44	44	___	44

Activities

There are four cacti.

Each cactus have eleven flowers.

How many flowers are there in all? _____

Use the numbers below to make equations.

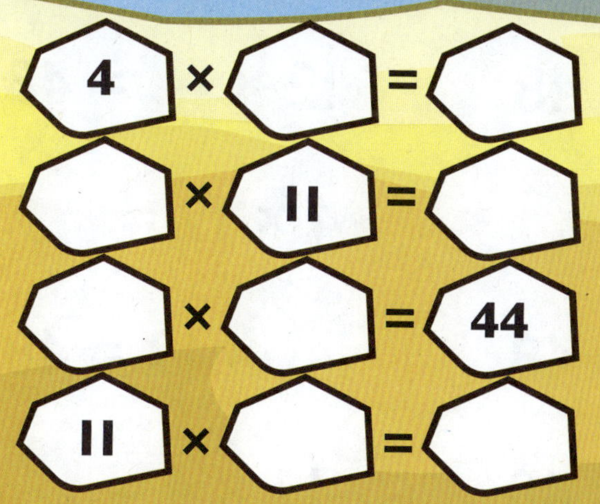

Solve the crossword puzzle.

4	×	11	=	44
×	■	×	■	■
	×		=	
=	■	=	■	■
44	■		■	■

Draw four ants crawling into each cactus.

Write the equation to match the picture. _____ × _____ = _____

Fill in the Answer

Fill in the blanks to complete the equations.

$4 \times 12 = \square$ $\square \times 12 = 48$ $12 \times \square = 48$

$12 \times \square = 48$ $12 \times 4 = \square$ $4 \times 12 = \square$

$\square \times 4 = 48$ $4 \times \square = 48$ $\square \times 12 = 48$

$4 \times \square = 48$ $\square \times 4 = 48$ $12 \times 4 = \square$

$\begin{array}{r} 4 \\ \times \square \\ \hline 48 \end{array}$ $\begin{array}{r} 4 \\ \times 12 \\ \hline \square \end{array}$ $\begin{array}{r} \square \\ \times 4 \\ \hline 48 \end{array}$ $\begin{array}{r} 12 \\ \times 4 \\ \hline \square \end{array}$ $\begin{array}{r} 4 \\ \times \square \\ \hline 48 \end{array}$ $\begin{array}{r} \square \\ \times 12 \\ \hline 48 \end{array}$

$\begin{array}{r} 12 \\ \times 4 \\ \hline \square \end{array}$ $\begin{array}{r} 4 \\ \times \square \\ \hline 48 \end{array}$ $\begin{array}{r} \square \\ \times 12 \\ \hline 48 \end{array}$ $\begin{array}{r} 4 \\ \times \square \\ \hline 48 \end{array}$ $\begin{array}{r} 4 \\ \times 12 \\ \hline \square \end{array}$ $\begin{array}{r} \square \\ \times 4 \\ \hline 48 \end{array}$

Activities

There are four rabbits. Each rabbit gets twelve carrots. How many carrots are there in all? _____

Use the numbers below to make equations.

4 12 48

There are 12 buttons. Each button has 4 holes.

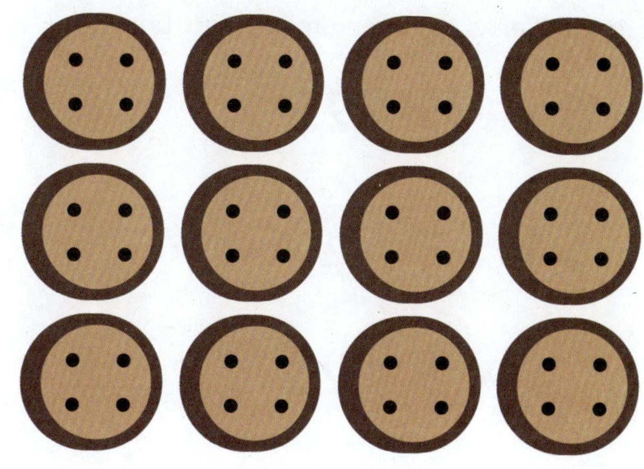

How many holes are there in all? _____

Draw four carrots for each rabbit.

Write the equation to match the picture.

_____ × _____ = _____

Fill in the Answer

Fill in the blanks to complete the equations.

5 × ___ = 25	5 × 5 = ___	5 × 5 = ___
5 × 5 = ___	___ × 5 = 25	___ × 5 = 25
___ × 5 = 25	5 × ___ = 25	5 × ___ = 25
5 × ___ = 25	5 × 5 = ___	___ × 5 = 25

5	5	___	5	5	___
× ___	× 5	× 5	× 5	× ___	× 5
25	___	25	___	25	25

5	5	___	5	5	___
× 5	× ___	× 5	× ___	× 5	× 5
___	25	25	25	___	25

Activities

There are five tow trucks.

Each tow truck tows five cars.

What is the total number of cars? _____

Use the numbers below to make equations.

There are 5 holes per die and 5 dice.

How many holes are there? _____

Draw five mud spots on each car.

Write the equation to match the picture.

_____ × _____ = _____

Fill in the Answer

Fill in the blanks to complete the equations.

$5 \times 6 = \square$ $6 \times 5 = \square$ $\square \times 6 = 30$

$6 \times \square = 30$ $\square \times 6 = 30$ $6 \times 5 = \square$

$\square \times 5 = 30$ $5 \times 6 = \square$ $\square \times 5 = 30$

$5 \times \square = 30$ $5 \times \square = 30$ $6 \times \square = 30$

$6 \times 5 = \square$ $\square \times 5 = 30$ $6 \times \square = 30$ $\square \times 5 = 30$ $5 \times \square = 30$ $6 \times 5 = \square$

$6 \times \square = 30$ $5 \times 6 = \square$ $\square \times 6 = 30$ $5 \times \square = 30$ $5 \times 6 = \square$ $\square \times 5 = 30$

Activities

There are five craters. Each crater has six rocks. What is the total number of rocks? ____

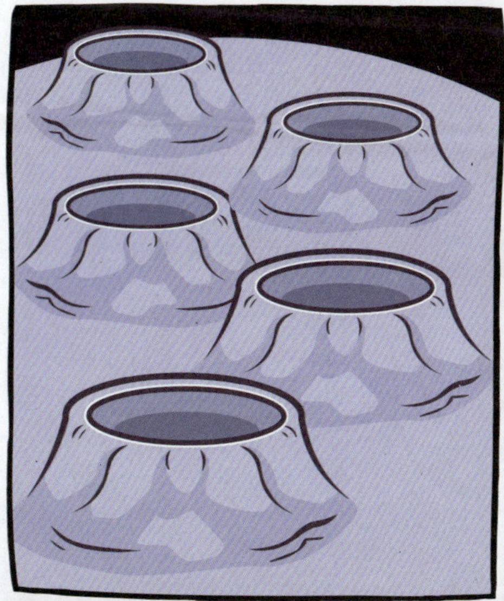

Use the numbers below to make equations.

5 6 30

5	×		=	
	×		=	30
	×	6	=	
	×	5	=	

Circle all of the 5, 6, 30 fact families.

5	6	30	5	3
6	9	5	20	35
30	5	3	30	21
6	2	5	6	30
3	6	5	30	1

Draw five planets around each sun.

Write the equation to match the picture. ______ × ______ = ______

Fill in the Answer

Fill in the blanks to complete the equations.

5 × 7 = ___	7 × ___ = 35	___ × 7 = 35
___ × 7 = 35	___ × 5 = 35	7 × ___ = 35
5 × ___ = 35	7 × 5 = ___	7 × 5 = ___
5 × 7 = ___	5 × ___ = 35	___ × 5 = 35

5	5	___	5	5	___
× ___	× 7	× 5	× 7	× ___	× 7
35	___	35	___	35	35

7	5	___	7	7	___
× 5	× ___	× 5	× ___	× 5	× 5
___	35	35	35	___	35

Activities

There are five mirrors.

Each mirror has seven light bulbs.

How many light bulbs are there in all?___

Use the numbers below to make equations.

5 7 35

5 × ___ = ___

___ × ___ = 35

7 × ___ = ___

___ × 5 = ___

Solve the crossword puzzle.

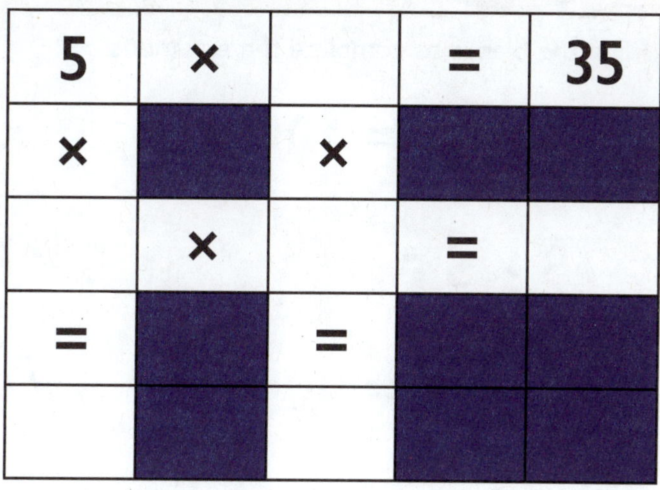

5	×		=	35
×		×		
	×		=	
=		=		

Draw five lights around each mirror.

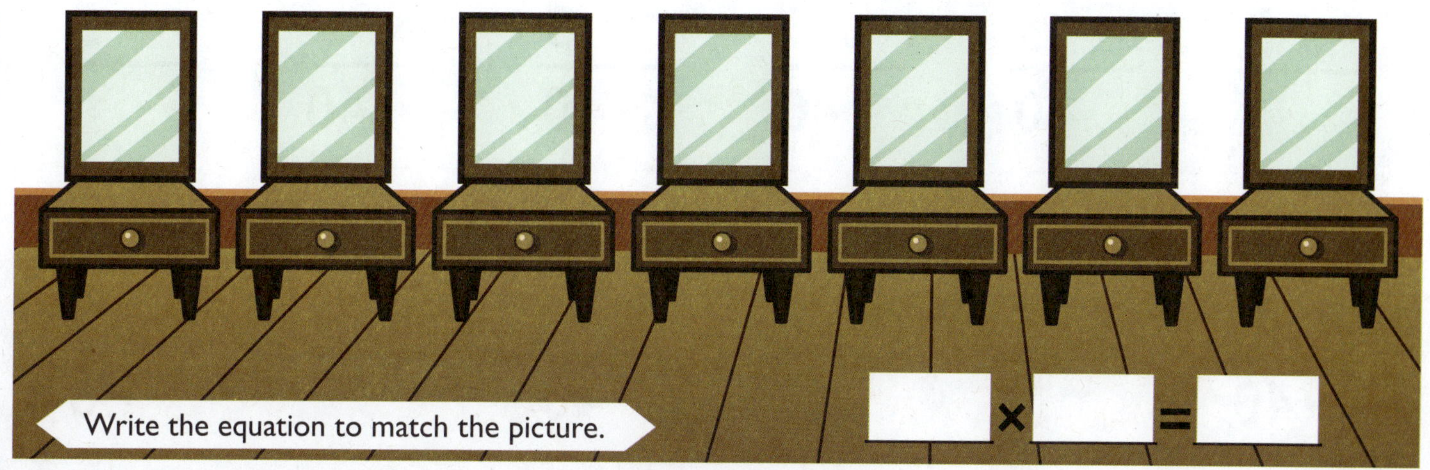

Write the equation to match the picture.

___ × ___ = ___

Fill in the Answer

Fill in the blanks to complete the equations.

5 × ___ = 40 | ___ × 5 = 40 | ___ × 5 = 40

8 × 5 = ___ | 5 × 8 = ___ | 8 × 5 = ___

___ × 8 = 40 | 8 × ___ = 40 | ___ × 8 = 40

8 × ___ = 40 | 5 × 8 = ___ | 5 × ___ = 40

8	___	8	___	5	8
× 5	× 5	× ___	× 5	× ___	× 5
___	40	40	40	40	___

8	5	___	8	5	___
× ___	× 8	× 8	× ___	× 8	× 5
40	___	40	40	___	40

Activities

There are five pizzas.

Each pizza has eight slices.

How many slices are there in total? ____

Use the numbers below to make equations.

5 8 40

$\bigcirc \times 5 = \bigcirc$

$8 \times \bigcirc = \bigcirc$

$\bigcirc \times \bigcirc = 40$

$5 \times \bigcirc = \bigcirc$

There are 5 holes per die and 8 dice.

How many holes are there? _____

Draw five gumballs in each gumball machine.

Write the equation to match the picture. _____ × _____ = _____

Fill in the Answer

Fill in the blanks to complete the equations.

___ × 5 = 45	5 × 9 = ___	___ × 5 = 45
9 × 5 = ___	5 × ___ = 45	5 × 9 = ___
5 × ___ = 45	___ × 9 = 45	___ × 9 = 45
9 × ___ = 45	9 × 5 = ___	9 × ___ = 45

9	5	___	5	5	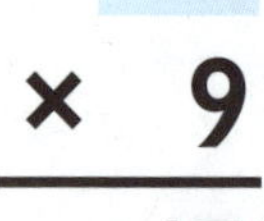
× ___	× 9	× 5	× 9	× ___	× 9
45	___	45	___	45	45

5	9	___	5	9	___
× 9	× ___	× 5	× ___	× 5	× 5
___	45	45	45	___	45

Activities

There are five cones. Each cone has nine rocks. How many rocks are there in total? ____

Use the numbers below to make equations.

There are 9 buttons. Each button has 5 holes.

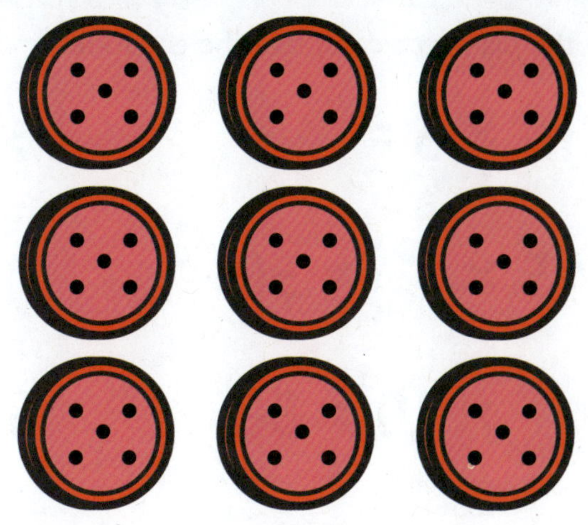

How many holes are there in all? _____

Draw five stripes on each cone.

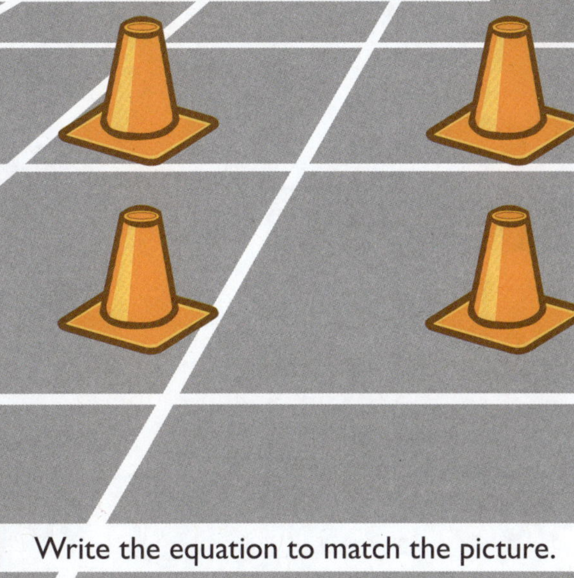

Write the equation to match the picture.

Fill in the Answer

Fill in the blanks to complete the equations.

$10 \times 5 = \square$	$10 \times \square = 50$	$5 \times 10 = \square$
$5 \times \square = 50$	$\square \times 5 = 50$	$10 \times \square = 50$
$\square \times 10 = 50$	$5 \times \square = 50$	$\square \times 5 = 50$
$5 \times 10 = \square$	$\square \times 10 = 50$	$10 \times 5 = \square$

$\begin{array}{r} 5 \\ \times\ 10 \\ \hline \square \end{array}$	$\begin{array}{r} \square \\ \times\ 5 \\ \hline 50 \end{array}$	$\begin{array}{r} 10 \\ \times\ \square \\ \hline 50 \end{array}$	$\begin{array}{r} \square \\ \times\ 5 \\ \hline 50 \end{array}$	$\begin{array}{r} 5 \\ \times\ \square \\ \hline 50 \end{array}$	$\begin{array}{r} 10 \\ \times\ 5 \\ \hline \square \end{array}$

$\begin{array}{r} 10 \\ \times\ \square \\ \hline 50 \end{array}$	$\begin{array}{r} 5 \\ \times\ 10 \\ \hline \square \end{array}$	$\begin{array}{r} \square \\ \times\ 10 \\ \hline 50 \end{array}$	$\begin{array}{r} 5 \\ \times\ \square \\ \hline 50 \end{array}$	$\begin{array}{r} 10 \\ \times\ 5 \\ \hline \square \end{array}$	$\begin{array}{r} \square \\ \times\ 5 \\ \hline 50 \end{array}$

Activities

There are five trees.

Each tree has ten branches.

How many branches are there in all? ____

Use the numbers below to make equations.

5 10 50

5	×		=	
	×	10	=	
	×	5	=	
	×		=	50

Circle all of the 5, 10, 50 fact families.

5	10	50	8	9
12	10	3	13	11
10	5	50	25	50
18	5	55	5	45
10	5	10	50	72

Draw ten spots on each giraffe.

Write the equation to match the picture.

_____ × _____ = _____

Fill in the Answer

Fill in the blanks to complete the equations.

11 × 5 = ___	___ × 5 = 55	___ × 5 = 55
5 × 11 = ___	11 × ___ = 55	5 × ___ = 55
5 × ___ = 55	___ × 11 = 55	11 × 5 = ___
___ × 11 = 55	11 × ___ = 55	5 × 11 = ___

11	5	___	5	5	___
× ___	× 11	× 5	× 11	× ___	× 11
55	___	55	___	55	55

5	11	___	11	11	___
× 11	× ___	× 5	× ___	× 5	× 5
___	55	55	55	___	55

Activities

There are five leis. Each lei has eleven flowers. How many flowers are there in all? ____

Use the numbers below to make equations.

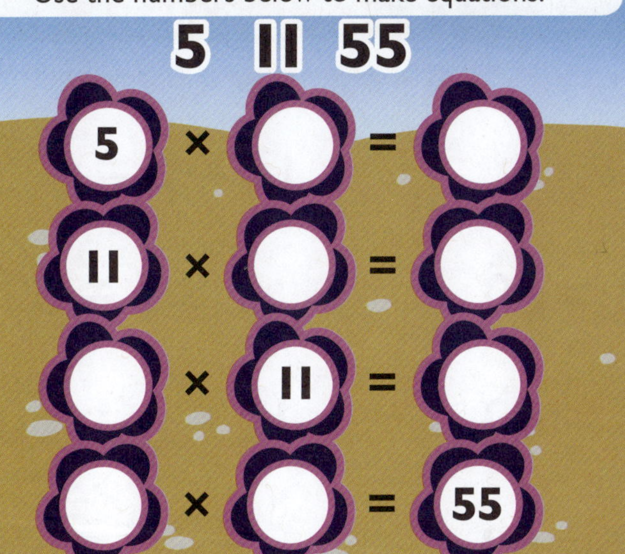

Solve the crossword puzzle.

5	×		=	
×	■	×	■	■
	×		=	
=	■	=	■	■
55	■		■	■

Draw five coconuts on each palm tree.

Write the equation to match the picture. ______ × ______ = ______

Fill in the Answer

Fill in the blanks to complete the equations.

____ × 5 = 60	12 × 5 = ____	12 × 5 = ____
5 × 12 = ____	____ × 12 = 60	____ × 5 = 60
12 × ____ = 60	5 × ____ = 60	12 × ____ = 60
5 × 12 = ____	____ × 12 = 60	5 × ____ = 60

5	____	12	____	5	12
× 12	× 5	× ____	× 5	× ____	× 5
____	60	60	60	60	____

12	5	____	5	12	____
× ____	× 12	× 12	× ____	× 5	× 5
60	____	60	60	____	60

Activities

There are twelve tow trucks. Each truck carries five cars. How many cars are there in all? ______

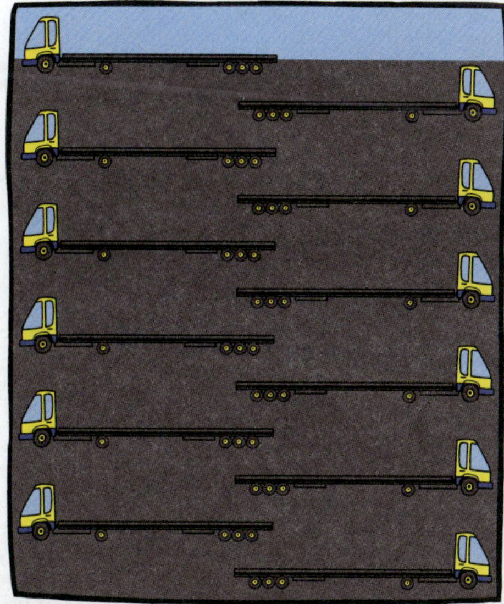

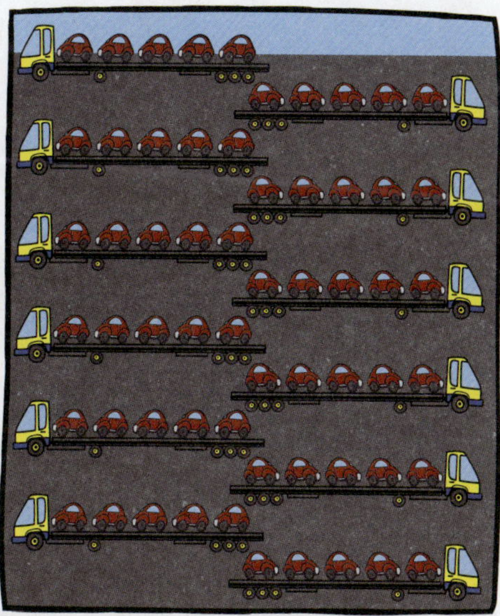

Use the numbers below to make equations.

5 12 60

5 × ◯ = ◯

12 × ◯ = ◯

◯ × ◯ = 60

◯ × 5 = ◯

There are 5 holes per die and 12 dice.

How many holes are there? ______

Draw five propellers on each helicopter.

Write the equation to match the picture. ______ × ______ = ______

Fill in the Answer

Fill in the blanks to complete the equations.

___ × 6 = 36	6 × 6 = ___	___ × 6 = 36
6 × 6 = ___	6 × ___ = 36	6 × 6 = ___
6 × ___ = 36	___ × 6 = 36	___ × 6 = 36
6 × ___ = 36	6 × 6 = ___	6 × ___ = 36

6	6	___	6	6	___
× ___	× 6	× 6	× 6	× ___	× 6
36	___	36	___	36	36

6	6	___	6	6	___
× 6	× ___	× 6	× ___	× 6	× 6
___	36	36	36	___	36

Activities

There are six hooks. There are six helmets per hook. How many helmets are there? ____

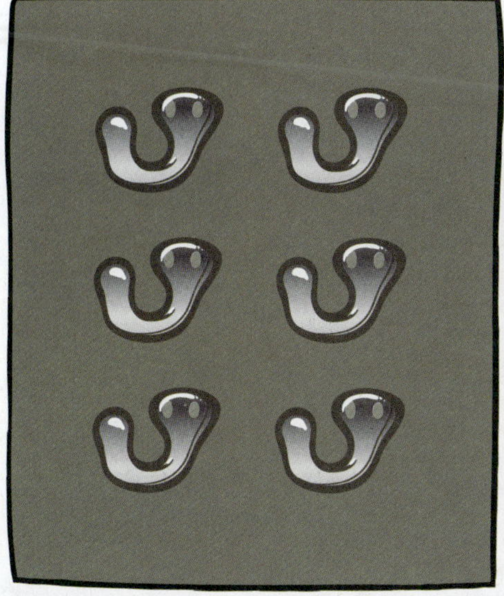

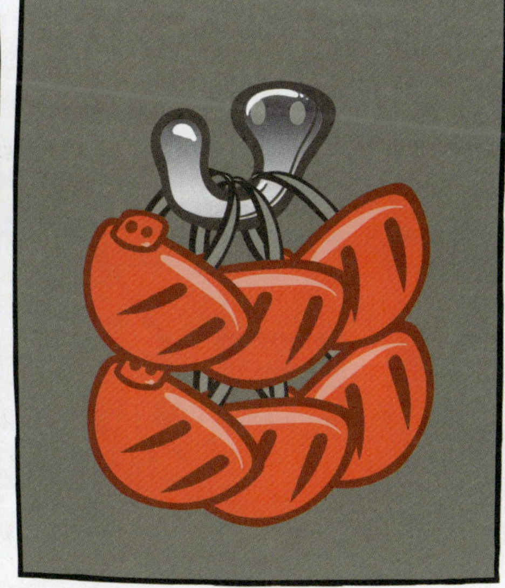

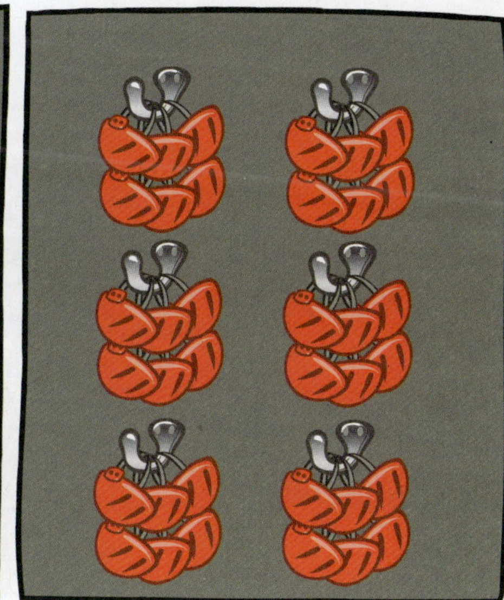

Use the numbers below to make equations.

There are 6 buttons. Each button has 6 holes.

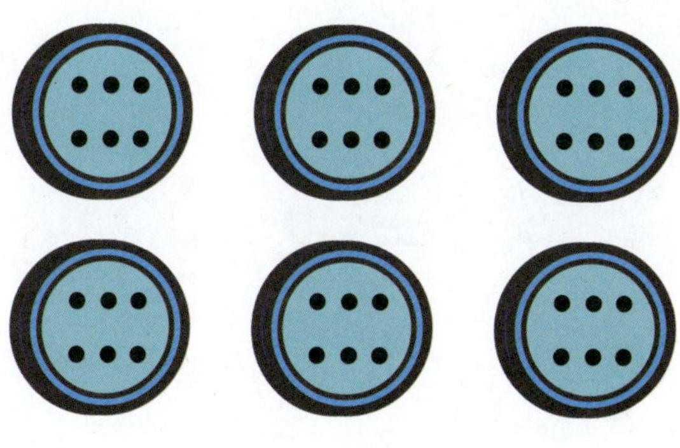

How many holes are there in all? _____

Draw six spokes on each tire.

Write the equation to match the picture. _____ × _____ = _____

Fill in the Answer

Fill in the blanks to complete the equations.

$7 \times \square = 42$ $\quad$ $7 \times 6 = \square$ $\quad$ $7 \times \square = 42$

$\square \times 7 = 42$ $\quad$ $\square \times 6 = 42$ $\quad$ $6 \times 7 = \square$

$6 \times 7 = \square$ $\quad$ $7 \times 6 = \square$ $\quad$ $6 \times \square = 42$

$\square \times 6 = 42$ $\quad$ $6 \times \square = 42$ $\quad$ $\square \times 7 = 42$

$\begin{array}{r} 6 \\ \times\ 7 \\ \hline \square \end{array}$ $\quad$ $\begin{array}{r} \square \\ \times\ 7 \\ \hline 42 \end{array}$ $\quad$ $\begin{array}{r} 6 \\ \times\ \square \\ \hline 42 \end{array}$ $\quad$ $\begin{array}{r} \square \\ \times\ 6 \\ \hline 42 \end{array}$ $\quad$ $\begin{array}{r} 7 \\ \times\ \square \\ \hline 42 \end{array}$ $\quad$ $\begin{array}{r} 7 \\ \times\ 6 \\ \hline \square \end{array}$

$\begin{array}{r} 7 \\ \times\ \square \\ \hline 42 \end{array}$ $\quad$ $\begin{array}{r} 6 \\ \times\ 7 \\ \hline \square \end{array}$ $\quad$ $\begin{array}{r} \square \\ \times\ 7 \\ \hline 42 \end{array}$ $\quad$ $\begin{array}{r} 6 \\ \times\ \square \\ \hline 42 \end{array}$ $\quad$ $\begin{array}{r} 7 \\ \times\ 6 \\ \hline \square \end{array}$ $\quad$ $\begin{array}{r} \square \\ \times\ 6 \\ \hline 42 \end{array}$

Activities

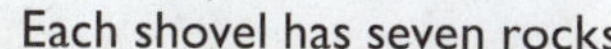

There are six shovels.

Each shovel has seven rocks.

How many rocks are there in all? ______

Use the numbers below to make equations.

6 7 42

6 × ___ = ___

7 × ___ = ___

___ × ___ = 42

___ × 6 = ___

Circle all of the 6, 7, 42 fact families.

6	7	42	5	3
7	9	6	20	35
42	5	3	42	21
6	2	7	6	42
3	6	5	42	1

Draw six boulders on each wheel barrel.

Write the equation to match the picture. ____ × ____ = ____

Fill in the Answer

Fill in the blanks to complete the equations.

8 × ___ = 48   6 × 8 = ___   6 × 8 = ___

___ × 8 = 48   6 × ___ = 48   8 × ___ = 48

6 × ___ = 48   ___ × 6 = 48   ___ × 8 = 48

___ × 6 = 48   8 × 6 = ___   8 × 6 = ___

---

6 × ___ = 48   8 × 6 = ___   ___ × 6 = 48   6 × 8 = ___   8 × ___ = 48   ___ × 6 = 48

---

8 × 6 = ___   6 × ___ = 48   ___ × 8 = 48   8 × ___ = 48   8 × 6 = ___   ___ × 6 = 48

Activities

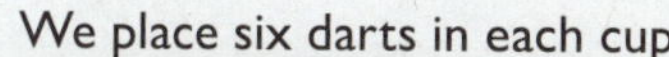

There are eight cups.

We place six darts in each cup.

How many darts are there in all? ______

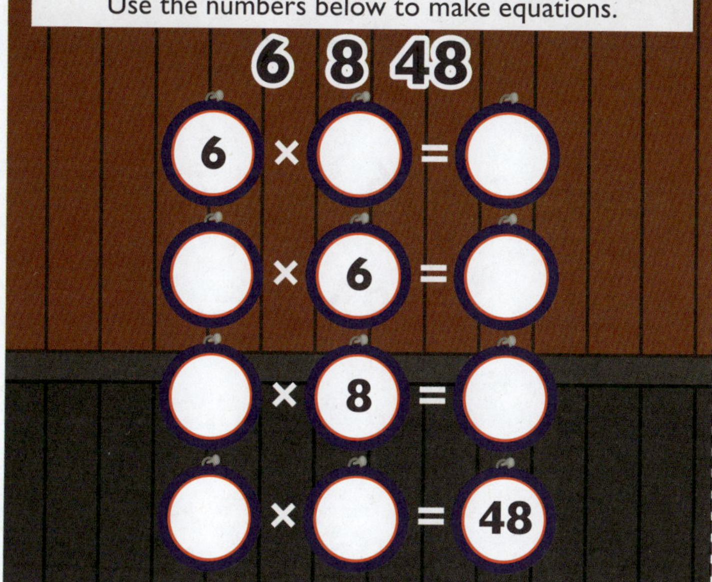

Solve the crossword puzzle.

6	×		=	
×		×		
	×		=	
=		=		
48				

Draw eight darts on each board.

Write the equation to match the picture. ______ × ______ = ______

Fill in the Answer

Fill in the blanks to complete the equations.

6 × 9 = ___	9 × ___ = 54	6 × ___ = 54
6 × ___ = 54	___ × 9 = 54	9 × 6 = ___
___ × 6 = 54	9 × 6 = ___	___ × 9 = 54
6 × 9 = ___	___ × 6 = 54	9 × ___ = 54

6	___	6	___	9	9
× 9	× 9	× ___	× 9	× ___	× 6
___	54	54	54	54	___

9	6	___	6	9	___
× ___	× 9	× 9	× ___	× 6	× 6
54	___	54	54	___	54

Activities

There are six stems.

Each stem grow nine pedals.

How many pedals are there in all? _____

Use the numbers below to make equations.

There are 6 holes per die and 9 dice.

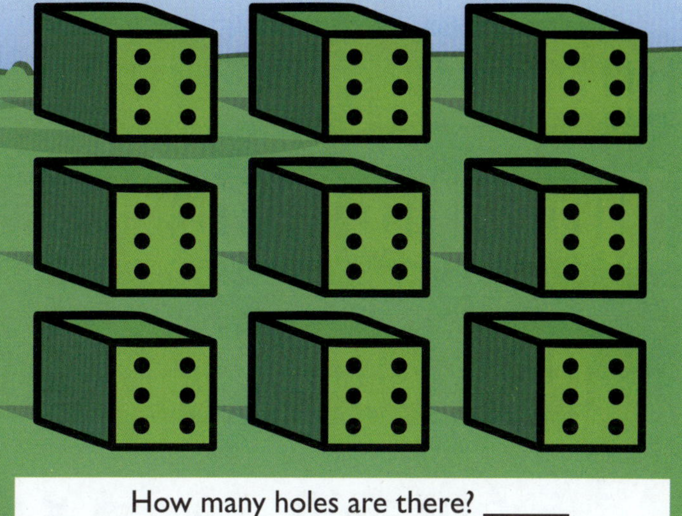

How many holes are there? _____

Draw six pedals on each stem.

Write the equation to match the picture.

_____ × _____ = _____

Fill in the Answer

Fill in the blanks to complete the equations.

10 × ___ = 60	___ × 6 = 60	10 × 6 = ___
6 × 10 = ___	6 × ___ = 60	6 × ___ = 60
10 × ___ = 60	___ × 10 = 60	___ × 6 = 60
___ × 10 = 60	6 × 10 = ___	10 × 6 = ___

6	10	___	6	10	___
× ___	× 6	× 6	× 10	× ___	× 6
60	___	60	___	60	60

10	6	___	6	10	___
× 6	× ___	× 10	× ___	× 6	× 6
___	60	60	60	___	60

Activities

There are six tunnels.

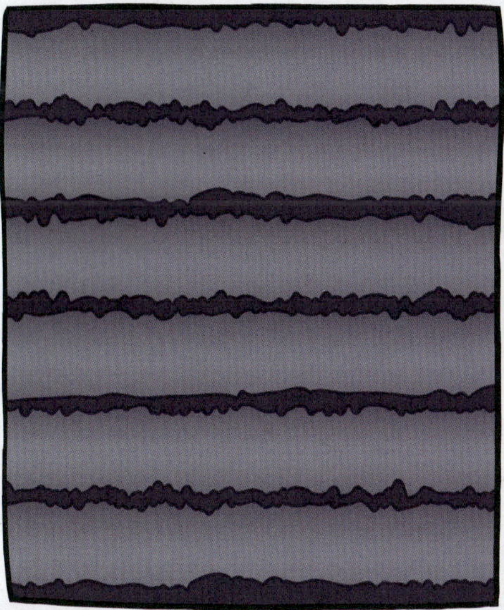

There are ten moles in each tunnel.

How many moles are there in all? _____

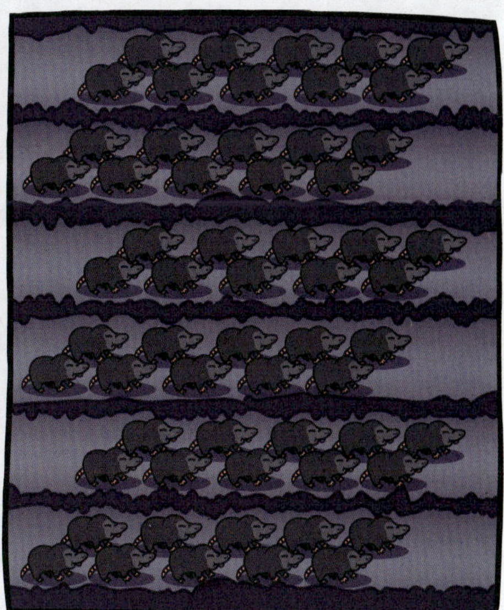

There are 10 buttons. Each button has 6 holes.

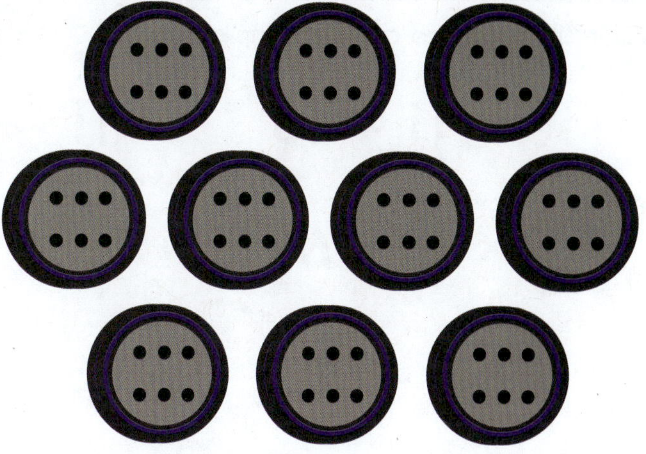

How many holes are there in all? _____

Draw ten grapes in each lunch box.

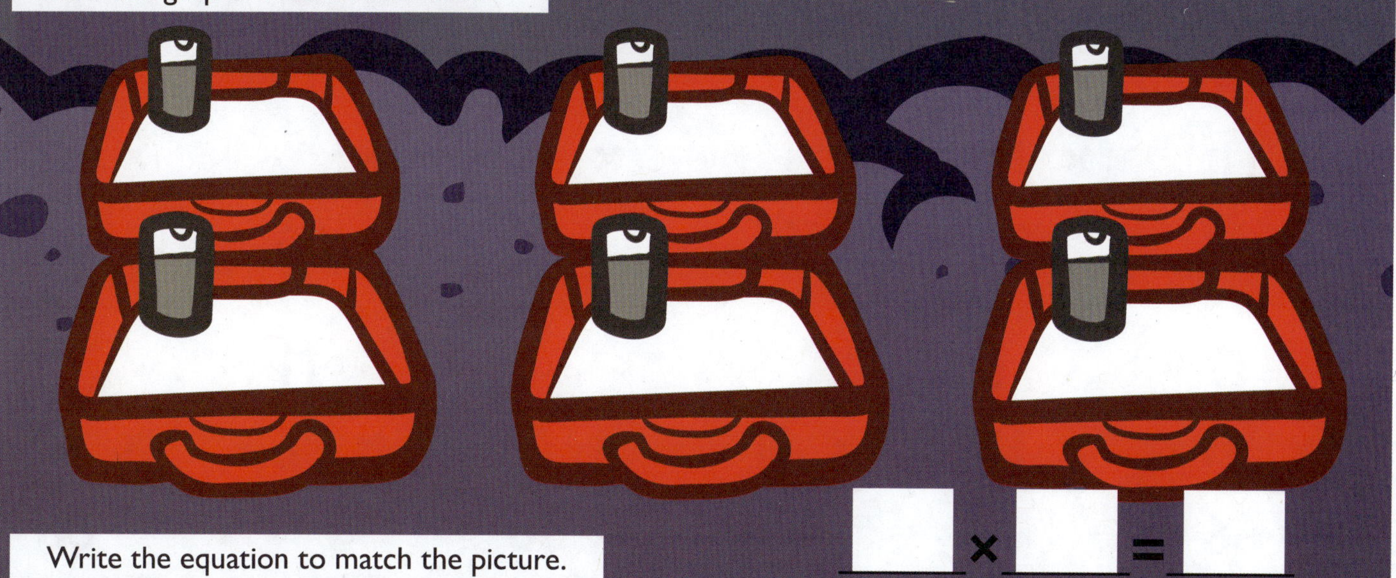

Write the equation to match the picture.

Fill in the Answer

Fill in the blanks to complete the equations.

___ × 11 = 66	11 × ___ = 66	11 × 6 = ___
6 × ___ = 66	___ × 11 = 66	11 × ___ = 66
11 × 6 = ___	6 × 11 = ___	___ × 6 = 66
___ × 6 = 66	6 × ___ = 66	6 × 11 = ___

6	___	6	___	11	11
× 11	× 11	× ___	× 11	× ___	× 6
___	66	66	66	66	___

11	6	___	6	11	___
× ___	× 11	× 11	× ___	× 6	× 6
66	___	66	66	___	66

Activities

There are six ladders. Each ladder has eleven steps. How many steps are there in all? ______

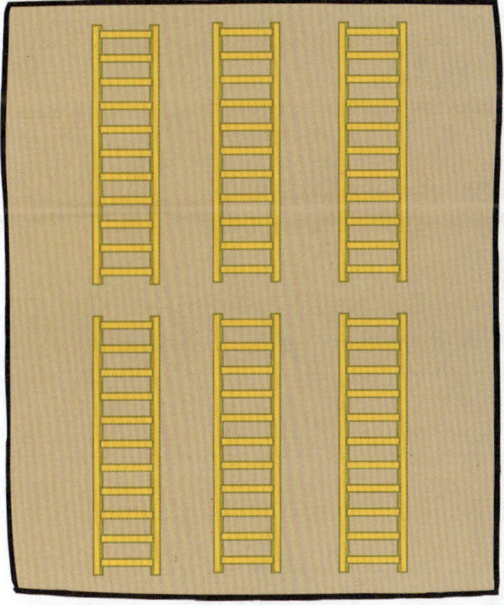

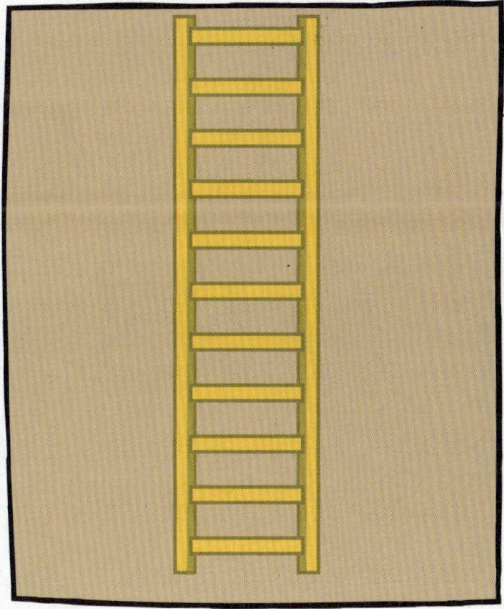

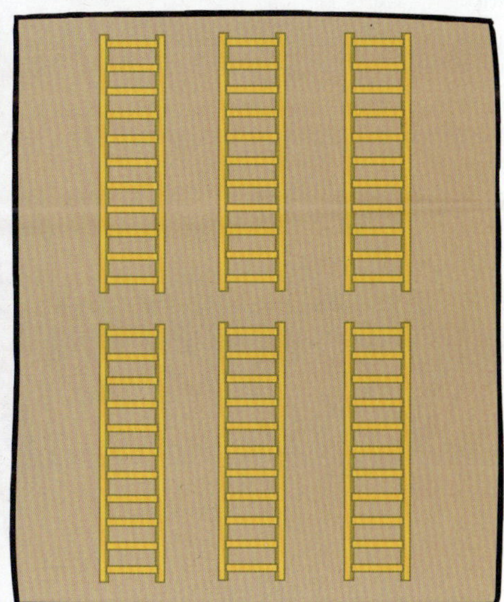

Use the numbers below to make equations.

6 11 66

Circle all of the 6, 11, 66 fact families.

11	6	66	11	5
6	9	11	6	66
66	5	3	66	30
6	11	66	6	30
3	6	11	66	1

Draw eleven keys on each key ring.

Write the equation to match the picture.

______ × ______ = ______

Fill in the Answer

Fill in the blanks to complete the equations.

12 × ___ = 72   12 × 6 = ___   ___ × 6 = 72

___ × 6 = 72   12 × ___ = 72   12 × 6 = ___

6 × 12 = ___   ___ × 12 = 72   ___ × 12 = 72

6 × ___ = 72   6 × 12 = ___   6 × ___ = 72

$$\begin{array}{r}6\\ \times\ \square\\ \hline 72\end{array}\quad \begin{array}{r}12\\ \times\ 6\\ \hline \square\end{array}\quad \begin{array}{r}\square\\ \times\ 6\\ \hline 72\end{array}\quad \begin{array}{r}6\\ \times\ 12\\ \hline \square\end{array}\quad \begin{array}{r}12\\ \times\ \square\\ \hline 72\end{array}\quad \begin{array}{r}\square\\ \times\ 6\\ \hline 72\end{array}$$

$$\begin{array}{r}12\\ \times\ 6\\ \hline \square\end{array}\quad \begin{array}{r}6\\ \times\ \square\\ \hline 72\end{array}\quad \begin{array}{r}\square\\ \times\ 12\\ \hline 72\end{array}\quad \begin{array}{r}6\\ \times\ \square\\ \hline 72\end{array}\quad \begin{array}{r}12\\ \times\ 6\\ \hline \square\end{array}\quad \begin{array}{r}\square\\ \times\ 6\\ \hline 72\end{array}$$

Activities

There are six tents.

Each tent gets twelve canteens.

How many canteens are there in all? ____

Use the numbers below to make equations.

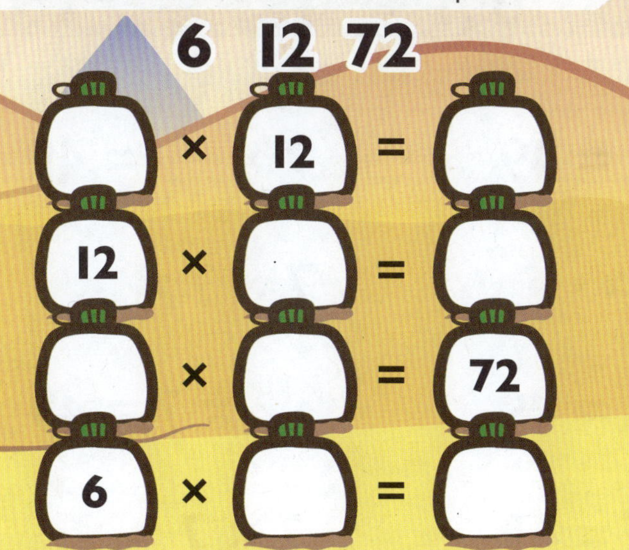

Solve the crossword puzzle.

6	×		=	
×	■	×	■	■
	×		=	
=	■	=	■	■
72	■		■	■

Draw twelve buttons on each vest.

Write the equation to match the picture. ______ × ______ = ______

Fill in the Answer

Fill in the blanks to complete the equations.

7 × 7 = ___	___ × 7 = 49	7 × ___ = 49
7 × ___ = 49	7 × 7 = ___	7 × 7 = ___
___ × 7 = 49	7 × ___ = 49	___ × 7 = 49
7 × ___ = 49	___ × 7 = 49	7 × 7 = ___

---

7 × 7 ___	___ × 7 49	7 × ___ 49	___ × 7 49	7 × ___ 49	7 × 7 ___

---

7 × ___ 49	7 × 7 ___	___ × 7 49	7 × ___ 49	7 × 7 ___	___ × 7 49

Activities

There are seven birds.

Each bird has seven seeds.

How many seeds are there in all? _____

Use the numbers below to make equations.

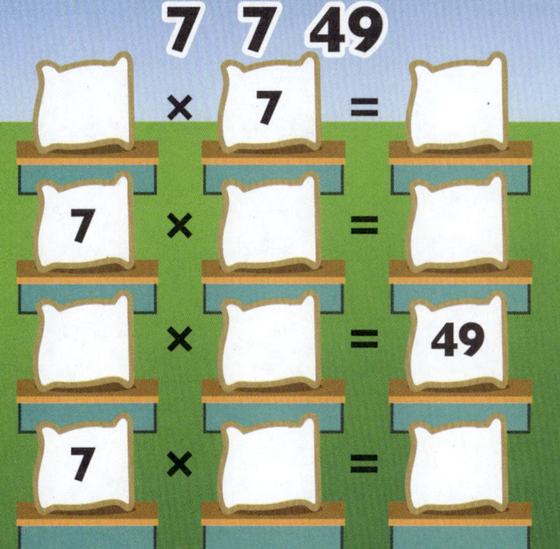

There are 7 nests and 7 eggs in each.

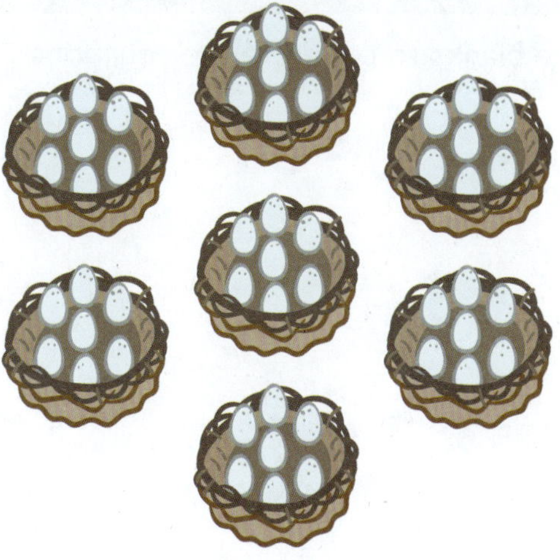

How many eggs are there in all? _____

Fill in the Answer

Fill in the blanks to complete the equations.

7 × ___ = 56	8 × 7 = ___	8 × 7 = ___
7 × 8 = ___	___ × 8 = 56	___ × 7 = 56
___ × 7 = 56	7 × ___ = 56	8 × ___ = 56
8 × ___ = 56	7 × 8 = ___	___ × 8 = 56

7	8	___	7	8	___
× ___	× 7	× 7	× 8	× ___	× 7
56	___	56	___	56	56

8	7	___	7	8	___
× 7	× ___	× 8	× ___	× 7	× 8
___	56	56	56	___	56

Activities

There are seven birds.

Each bird carries eight rags.

How many rags are there in all? _____

There are 8 buttons. Each button has 7 holes.

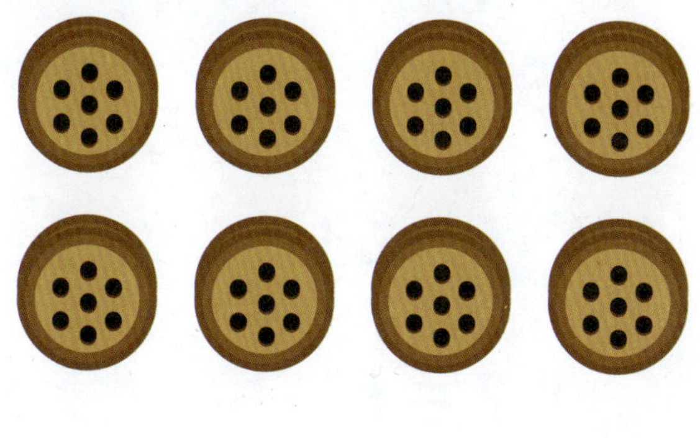

How many holes are there in all? _____

Draw eight pins on each clothesline.

Write the equation to match the picture. _____ × _____ = _____

Fill in the Answer

Fill in the blanks to complete the equations.

7 × 9 = ____	9 × 7 = ____	____ × 9 = 63
9 × ____ = 63	____ × 9 = 63	9 × 7 = ____
____ × 7 = 63	7 × 9 = ____	____ × 7 = 63
7 × ____ = 63	7 × ____ = 63	9 × ____ = 63

7	____	7	____	7	9
× 9	× 7	× ____	× 9	× ____	× 7
____	63	63	63	63	____

9	9	____	9	7	____
× ____	× 7	× 7	× ____	× 9	× 7
63	____	63	63	____	63

Activities

There are nine hamsters.

Each hamster get seven food pellets.

How many pellets are there in all? _____

Use the numbers below to make equations.

7 9 63

◯ × 7 = ◯

9 × ◯ = ◯

◯ × ◯ = 63

7 × ◯ = ◯

Circle all of the 7, 9, 63 fact families.

7	9	63	8	9
12	9	3	13	11
9	7	63	25	63
18	7	55	7	45
9	7	9	63	72

Draw nine hamsters on each tunnel.

Write the equation to match the picture.

_____ × _____ = _____

Fill in the Answer

Fill in the blanks to complete the equations.

7 × 10 = ____	10 × ____ = 70	____ × 10 = 70
____ × 7 = 70	____ × 10 = 70	7 × ____ = 70
7 × ____ = 70	7 × 10 = ____	10 × 7 = ____
10 × 7 = ____	10 × ____ = 70	____ × 7 = 70

7	10	____	7	10	____
× ____	× 7	× 7	× 10	× ____	× 7
70	____	70	____	70	70

10	7	____	7	10	____
× 7	× ____	× 10	× ____	× 7	× 10
____	70	70	70	____	70

Activities

There seven posts.

Each post has ten strips.

How many strips are there in all? _____

Use the numbers below to make equations.

Solve the crossword puzzle.

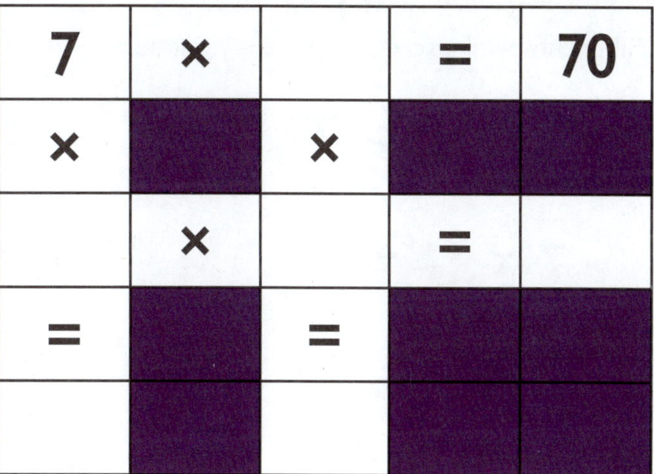

7	×		=	70
×		×		
	×		=	
=		=		

Draw seven brushes in each bucket.

Write the equation to match the picture.

_____ × _____ = _____

Fill in the Answer

Fill in the blanks to complete the equations.

$11 \times \square = 77$ $\quad \square \times 11 = 77$ $\quad \square \times 11 = 77$

$7 \times 11 = \square$ $\quad 11 \times 7 = \square$ $\quad 7 \times 11 = \square$

$\square \times 7 = 77$ $\quad 7 \times \square = 77$ $\quad \square \times 7 = 77$

$7 \times \square = 77$ $\quad 11 \times 7 = \square$ $\quad 11 \times \square = 77$

$$\begin{array}{r} 7 \\ \times\ 11 \\ \hline \square \end{array} \quad \begin{array}{r} \square \\ \times\ 7 \\ \hline 77 \end{array} \quad \begin{array}{r} 7 \\ \times\ \square \\ \hline 77 \end{array} \quad \begin{array}{r} \square \\ \times\ 11 \\ \hline 77 \end{array} \quad \begin{array}{r} 7 \\ \times\ \square \\ \hline 77 \end{array} \quad \begin{array}{r} 11 \\ \times\ 7 \\ \hline \square \end{array}$$

$$\begin{array}{r} 11 \\ \times\ \square \\ \hline 77 \end{array} \quad \begin{array}{r} 11 \\ \times\ 7 \\ \hline \square \end{array} \quad \begin{array}{r} \square \\ \times\ 7 \\ \hline 77 \end{array} \quad \begin{array}{r} 11 \\ \times\ \square \\ \hline 77 \end{array} \quad \begin{array}{r} 7 \\ \times\ 11 \\ \hline \square \end{array} \quad \begin{array}{r} \square \\ \times\ 7 \\ \hline 77 \end{array}$$

Activities

There are seven buckets. Each bucket has eleven drumsticks. How many drumsticks are there in all? _____

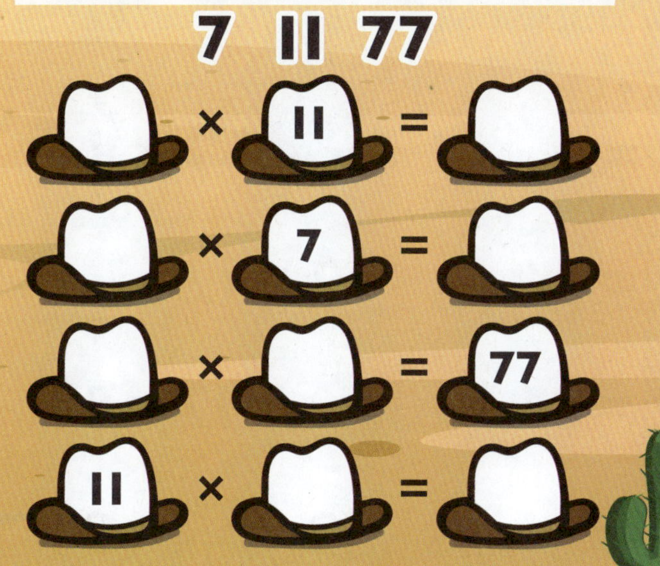

There are 11 xylophones and 7 plates in each.

How many plates are there in all? _____

Draw eleven mallets in each bucket.

Fill in the Answer

Fill in the blanks to complete the equations.

☐ × 7 = 84	7 × 12 = ☐	☐ × 7 = 84
12 × 7 = ☐	7 × ☐ = 84	7 × 12 = ☐
7 × ☐ = 84	☐ × 12 = 84	☐ × 12 = 84
12 × ☐ = 84	12 × 7 = ☐	12 × ☐ = 84

---

7 × ☐ = 84	12 × 7 = ☐	☐ × 7 = 84	7 × 12 = ☐	12 × ☐ = 84	☐ × 7 = 84

---

12 × 7 = ☐	7 × ☐ = 84	☐ × 12 = 84	7 × ☐ = 84	12 × 7 = ☐	☐ × 12 = 84

Activities

There are seven dog bowls.

Each bowl has twelve pieces of dog food.

How many pieces are there in all? _____

Use the numbers below to make equations.

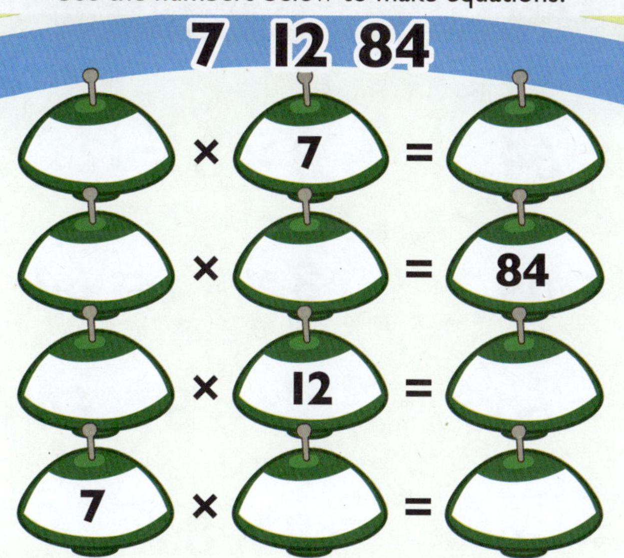

There are 12 buttons. Each button has 7 holes.

How many holes are there in all? _____

Draw seven buttons on each control panel.

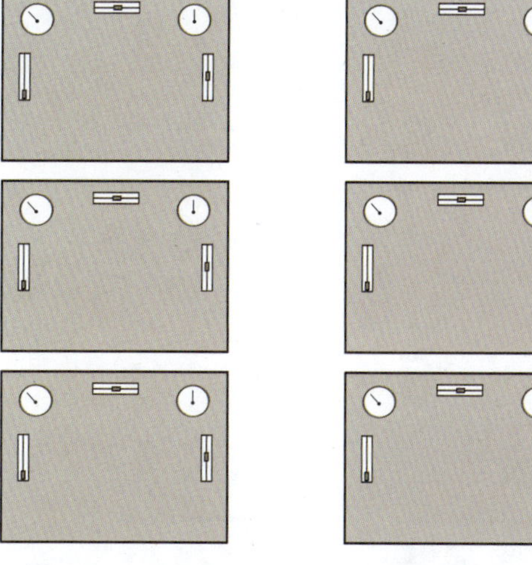

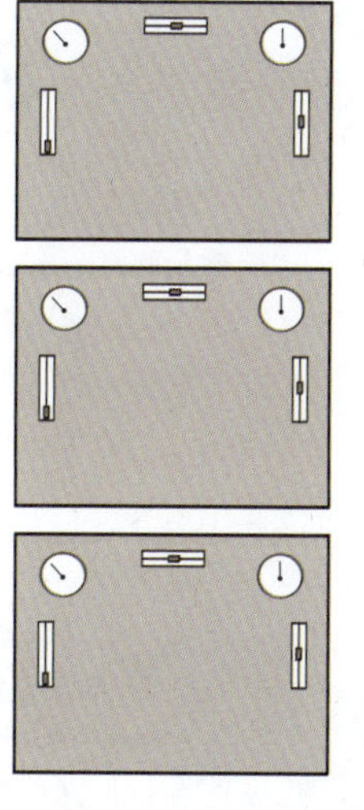

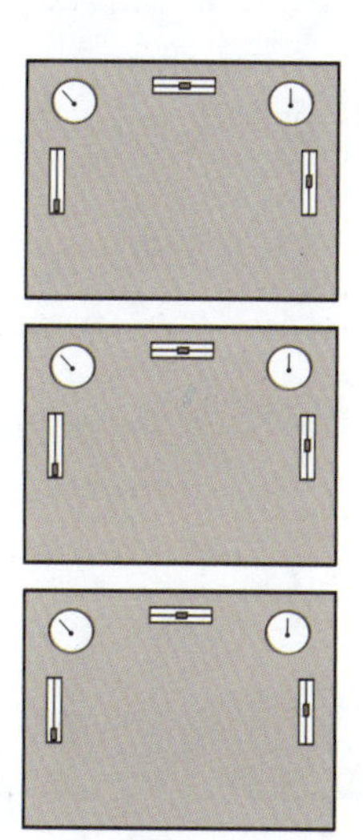

Write the equation to match the picture. _______ × _______ = _______

Fill in the Answer

Fill in the blanks to complete the equations.

8 × 8 = ___	8 × ___ = 64	8 × 8 = ___
8 × ___ = 64	___ × 8 = 64	8 × ___ = 64
___ × 8 = 64	8 × ___ = 64	___ × 8 = 64
8 × 8 = ___	___ × 8 = 64	8 × 8 = ___

8	___	8	___	8	8
× 8	× 8	× ___	× 8	× ___	× 8
___	64	64	64	64	___

8	8	___	8	8	___
× ___	× 8	× 8	× ___	× 8	× 8
64	___	64	64	___	64

Activities

There are eight kids.

Each kid gets eight awards.

How many awards are there in all? _____

Use the numbers below to make equations.

Circle all of the 8, 8, 64 fact families.

8	8	64	9	8
8	12	3	13	11
10	8	50	25	64
18	4	64	8	45
12	8	8	64	72

Draw eight planets on each mobile hanger.

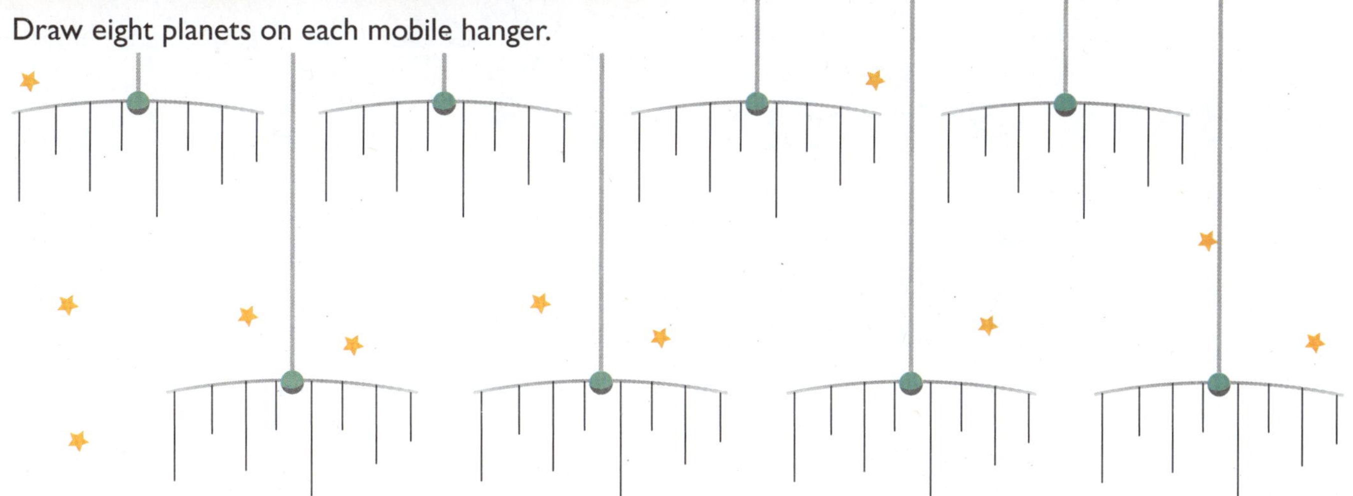

Write the equation to match the picture. _____ × _____ = _____

Fill in the Answer

Fill in the blanks to complete the equations.

8 × 9 = ___	___ × 9 = 72	___ × 9 = 72
9 × 8 = ___	8 × ___ = 72	8 × ___ = 72
9 × ___ = 72	___ × 8 = 72	9 × 8 = ___
___ × 8 = 72	9 × ___ = 72	8 × 9 = ___

8	9	___	8	9	___
× ___	× 8	× 9	× 9	× ___	× 8
72	___	72	___	72	72

9	8	___	8	9	___
× 8	× ___	× 9	× ___	× 8	× 8
___	72	72	72	___	72

Activities

There are nine hats.

Each hat has eight tassels.

How many tassels are there in all? ______

Use the numbers below to make equations.

8 9 72

8 × ◯ = ◯

◯ × ◯ = 72

◯ × 8 = ◯

9 × ◯ = ◯

Solve the crossword puzzle.

8	×		=	
×		×		
	×	8	=	
=		=		
72				

Draw eight coins in each purse.

Write the equation to match the picture.

______ × ______ = ______

Fill in the Answer

Fill in the blanks to complete the equations.

___ × 8 = 80	10 × 8 = ___	10 × 8 = ___
8 × 10 = ___	___ × 10 = 80	___ × 8 = 80
10 × ___ = 80	8 × ___ = 80	10 × ___ = 80
8 × 10 = ___	___ × 10 = 80	8 × ___ = 80

8	___	10	___	8	10
× 10	× 8	× ___	× 8	× ___	× 8
___	80	80	80	80	___

10	8	___	10	8	___
× ___	× 10	× 8	× ___	× 10	× 8
80	___	80	80	___	80

Activities

There are ten key rings. Each key ring has eight keys. How many keys are there in all? _____

Use the numbers below to make equations.

8 10 80

8	×		=	
	×		=	80
	×	10	=	
10	×		=	

There are 10 bunches with 8 bananas in each.

How many bananas in all? _____

Draw ten bananas in each box.

Write the equation to match the picture. _____ × _____ = _____

Fill in the Answer

Fill in the blanks to complete the equations.

___ × 8 = 88	11 × 8 = ___	___ × 8 = 88
8 × 11 = ___	8 × ___ = 88	11 × 8 = ___
11 × ___ = 88	___ × 11 = 88	___ × 11 = 88
8 × ___ = 88	8 × 11 = ___	11 × ___ = 88

8	11	___	8	11	___
× ___	× 8	× 8	× 11	× ___	× 8
88	___	88	___	88	88

11	8	___	8	11	___
× 8	× ___	× 11	× ___	× 8	× 11
___	88	88	88	___	88

Activities

There are eight kegs.

There are eleven holes on each keg.

How many holes are there in all? _____

Use the numbers below to make equations.

There are 11 buttons. Each button has 8 holes.

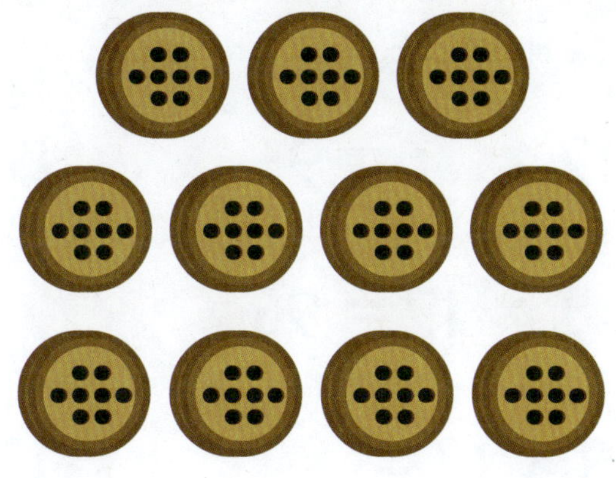

How many holes are there in all? _____

Draw eight barbs on each wire.

Write the equation to match the picture. _____ × _____ = _____

Fill in the Answer

Fill in the blanks to complete the equations.

12 × ___ = 96	12 × 8 = ___	12 × ___ = 96
___ × 12 = 96	___ × 8 = 96	8 × 12 = ___
8 × 12 = ___	12 × 8 = ___	8 × ___ = 96
___ × 8 = 96	8 × ___ = 96	___ × 12 = 96

$$\begin{array}{r} 8 \\ \times\ 12 \\ \hline \square \end{array} \quad \begin{array}{r} \square \\ \times\ 8 \\ \hline 96 \end{array} \quad \begin{array}{r} 12 \\ \times\ \square \\ \hline 96 \end{array} \quad \begin{array}{r} \square \\ \times\ 12 \\ \hline 96 \end{array} \quad \begin{array}{r} 8 \\ \times\ \square \\ \hline 96 \end{array} \quad \begin{array}{r} 12 \\ \times\ 8 \\ \hline \square \end{array}$$

$$\begin{array}{r} 12 \\ \times\ \square \\ \hline 96 \end{array} \quad \begin{array}{r} 8 \\ \times\ 12 \\ \hline \square \end{array} \quad \begin{array}{r} \square \\ \times\ 8 \\ \hline 96 \end{array} \quad \begin{array}{r} 8 \\ \times\ \square \\ \hline 96 \end{array} \quad \begin{array}{r} 12 \\ \times\ 8 \\ \hline \square \end{array} \quad \begin{array}{r} \square \\ \times\ 12 \\ \hline 96 \end{array}$$

Activities

There are eight igloos. Each igloo is made from twelve ice cubes. How many ice cubes are there in all? _____

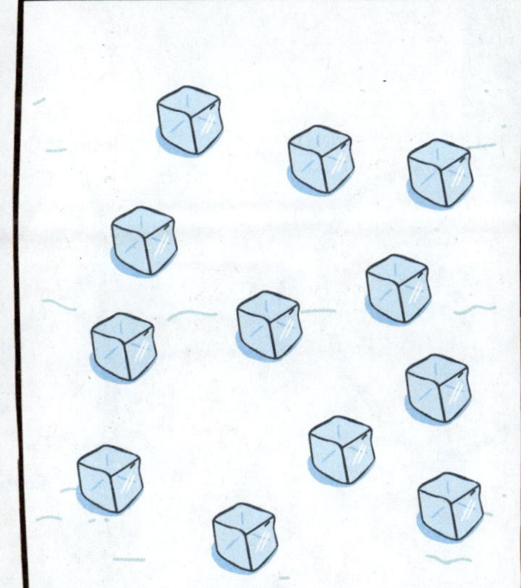

Use the numbers below to make equations.

8 12 96

8	×		=	
12	×		=	
	×	8	=	
	×		=	96

Circle all of the 8, 12, 96 fact families.

8	12	96	8	9
12	12	3	13	11
10	8	50	25	96
18	4	96	8	45
12	8	12	96	72

Draw eight snowballs for each baby monkey.

Write the equation to match the picture. _______ × _______ = _______

Fill in the Answer

Fill in the blanks to complete the equations.

$9 \times \square = 81$ $\quad 9 \times 9 = \square$ $\quad 9 \times 9 = \square$

$\square \times 9 = 81$ $\quad 9 \times \square = 81$ $\quad 9 \times \square = 81$

$9 \times \square = 81$ $\quad \square \times 9 = 81$ $\quad \square \times 9 = 81$

$\square \times 9 = 81$ $\quad 9 \times 9 = \square$ $\quad 9 \times 9 = \square$

$$\begin{array}{r} 9 \\ \times\ \square \\ \hline 81 \end{array} \quad \begin{array}{r} 9 \\ \times\ 9 \\ \hline \square \end{array} \quad \begin{array}{r} \square \\ \times\ 9 \\ \hline 81 \end{array} \quad \begin{array}{r} 9 \\ \times\ 9 \\ \hline \square \end{array} \quad \begin{array}{r} 9 \\ \times\ \square \\ \hline 81 \end{array} \quad \begin{array}{r} \square \\ \times\ 9 \\ \hline 81 \end{array}$$

$$\begin{array}{r} 9 \\ \times\ 9 \\ \hline \square \end{array} \quad \begin{array}{r} 9 \\ \times\ \square \\ \hline 81 \end{array} \quad \begin{array}{r} \square \\ \times\ 9 \\ \hline 81 \end{array} \quad \begin{array}{r} 9 \\ \times\ \square \\ \hline 81 \end{array} \quad \begin{array}{r} 9 \\ \times\ 9 \\ \hline \square \end{array} \quad \begin{array}{r} \square \\ \times\ 9 \\ \hline 81 \end{array}$$

Activities

There are nine palm trees.

Each palm tree has nine coconuts.

How many coconuts are there in all? ___

Use the numbers below to make equations.

9 9 81

9	×		=	
	×		=	81
	×	9	=	
9	×		=	

Solve the crossword puzzle.

9	×	9	=	
×		×		
9	×		=	81
=		=		

Draw nine coconuts in each basket.

Write the equation to match the picture. ______ × ______ = ______

Fill in the Answer

Fill in the blanks to complete the equations.

9 × 10 = ___	10 × ___ = 90	9 × ___ = 90
9 × ___ = 90	___ × 10 = 90	10 × 9 = ___
___ × 9 = 90	10 × 9 = ___	___ × 10 = 90
9 × 10 = ___	___ × 9 = 90	10 × ___ = 90

9	___	10	___	9	10
× 10	× 9	× ___	× 10	× ___	× 9
___	90	90	90	90	___

10	9	___	9	10	___
× ___	× 10	× 10	× ___	× 9	× 9
90	___	90	90	___	90

Activities

There are ten bubble wands. Each bubble wand blows nine bubbles. How many bubbles are there in all? ____

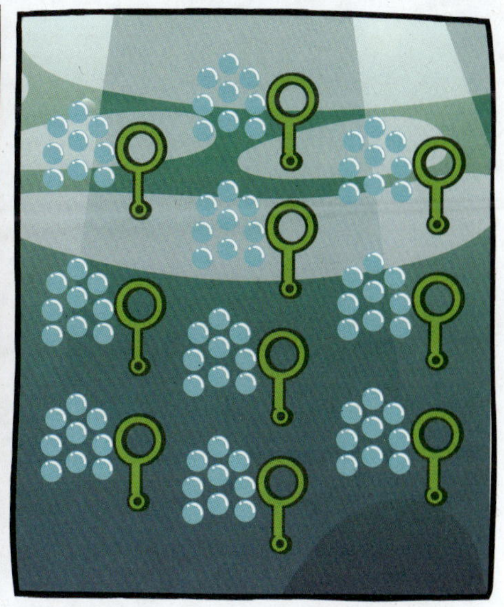

Use the numbers below to make equations.

9 10 90

10 × ___ = ___

___ × 10 = ___

___ × ___ = 90

9 × ___ = ___

There are 9 clams and 10 pearls in each clam.

How many pearls are there in all? _____

Draw nine spots on each turtle.

Write the equation to match the picture. ______ × ______ = ______

Fill in the Answer

Fill in the blanks to complete the equations.

11 × ___ = 99	___ × 9 = 99	11 × 9 = ___
9 × 11 = ___	9 × ___ = 99	9 × ___ = 99
11 × ___ = 99	___ × 11 = 99	___ × 9 = 99
___ × 11 = 99	9 × 11 = ___	11 × 9 = ___

9	___	11	___	9	11
× 11	× 9	× ___	× 9	× ___	× 9
___	99	99	99	99	___

9	9	___	11	9	___
× ___	× 11	× 9	× ___	× 11	× 9
99	___	99	99	___	99

Activities

There are nine microscopes.

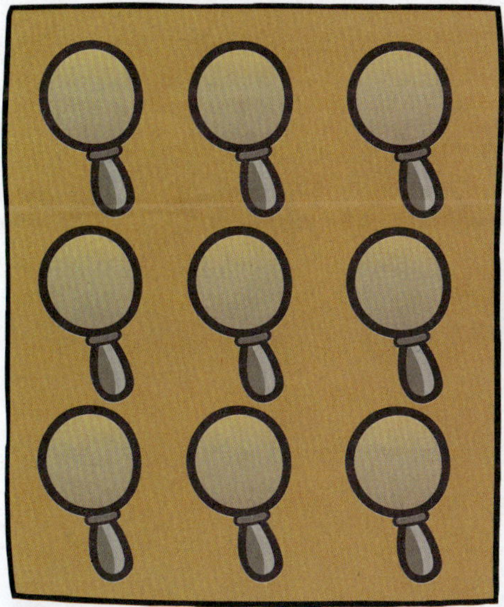

Each microscope finds eleven paws.

How many paws are there in all? _____

Use the numbers below to make equations.

9 11 99

There are 11 buttons. Each button has 9 holes.

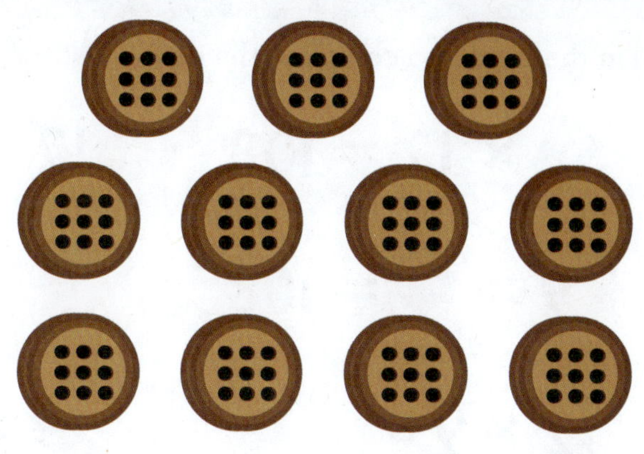

How many holes are there in all? _____

Draw eleven bones in each hole.

Write the equation to match the picture.

_____ × _____ = _____

Fill in the Answer

Fill in the blanks to complete the equations.

$\_\_\_ \times 12 = 108$ $\quad 12 \times \_\_\_ = 108$ $\quad 12 \times 9 = \_\_\_$

$9 \times \_\_\_ = 108$ $\quad \_\_\_ \times 12 = 108$ $\quad 12 \times \_\_\_ = 108$

$12 \times 9 = \_\_\_$ $\quad 9 \times 12 = \_\_\_$ $\quad \_\_\_ \times 9 = 108$

$\_\_\_ \times 9 = 108$ $\quad 9 \times \_\_\_ = 108$ $\quad 9 \times 12 = \_\_\_$

$$\begin{array}{r} 9 \\ \times\ 12 \\ \hline \_\_\_ \end{array} \quad \begin{array}{r} \_\_\_ \\ \times\ 9 \\ \hline 108 \end{array} \quad \begin{array}{r} 12 \\ \times\ \_\_\_ \\ \hline 108 \end{array} \quad \begin{array}{r} \_\_\_ \\ \times\ 9 \\ \hline 108 \end{array} \quad \begin{array}{r} 9 \\ \times\ \_\_\_ \\ \hline 108 \end{array} \quad \begin{array}{r} 12 \\ \times\ 9 \\ \hline \_\_\_ \end{array}$$

$$\begin{array}{r} 12 \\ \times\ \_\_\_ \\ \hline 108 \end{array} \quad \begin{array}{r} 9 \\ \times\ 12 \\ \hline \_\_\_ \end{array} \quad \begin{array}{r} \_\_\_ \\ \times\ 9 \\ \hline 108 \end{array} \quad \begin{array}{r} 12 \\ \times\ \_\_\_ \\ \hline 108 \end{array} \quad \begin{array}{r} 9 \\ \times\ 12 \\ \hline \_\_\_ \end{array} \quad \begin{array}{r} \_\_\_ \\ \times\ 9 \\ \hline 108 \end{array}$$

Activities

There are twelve lines of rope.

Each rope has nine knots.

How many knots are there in all? ______

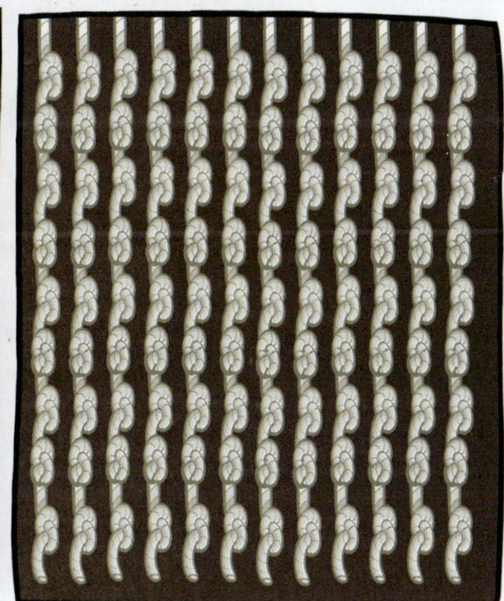

Use the numbers below to make equations.

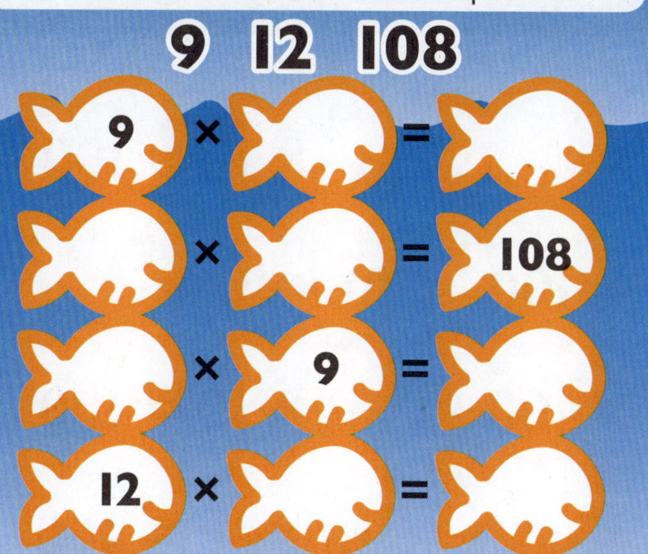

Circle all of the 9, 12, 108 fact families.

11	6	9	12	108
6	9	12	20	35
66	5	108	30	9
6	9	66	6	12
12	9	108	66	108

Draw twelve strips on each lighthouse.

Write the equation to match the picture. ______ × ______ = ______

Fill in the Answer

Fill in the blanks to complete the equations.

10 × ___ = 100   10 × 10 = ___   ___ × 10 = 100

___ × 10 = 100   10 × ___ = 100   10 × 10 = ___

10 × 10 = ___   ___ × 10 = 100   ___ × 10 = 100

10 × ___ = 100   10 × 10 = ___   10 × ___ = 100

10	___	10	___	10	10
× 10	× 10	× ___	× 10	× ___	× 10
___	100	100	100	100	___

10	10	___	10	10	___
× ___	× 10	× 10	× ___	× 10	× 10
100	___	100	100	___	100

Activities

There are ten hooks.

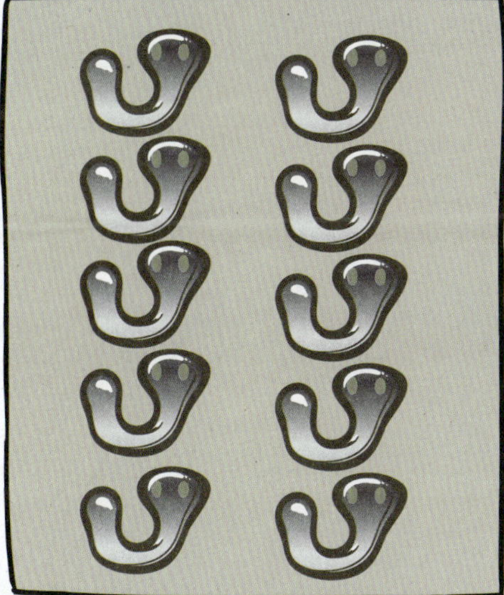

There are ten sets of goggles on each hook.

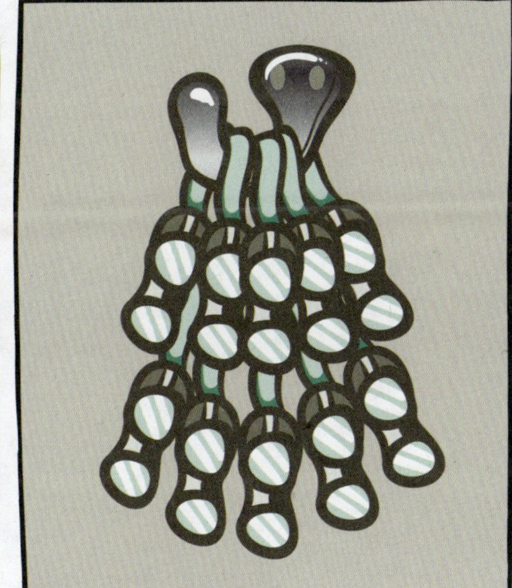

How many sets of goggles are there in all? _____

Solve the crossword puzzle.

10	×		=	
×		×		
	×	10	=	
=		=		
100				

Fill in the Answer

Fill in the blanks to complete the equations.

11 × 10 = ___	___ × 10 = 110	10 × ___ = 110
10 × ___ = 110	10 × 11 = ___	11 × 10 = ___
___ × 11 = 110	11 × ___ = 110	___ × 10 = 110
11 × ___ = 110	___ × 11 = 110	10 × 11 = ___

10 × 11 ___	___ × 10 110	11 × ___ 110	___ × 10 110	10 × ___ 110	11 × 10 ___
11 × ___ 110	10 × 11 ___	___ × 10 110	11 × ___ 110	10 × 11 ___	___ × 10 110

Activities

There are ten birds.

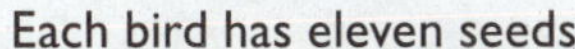

Each bird has eleven seeds.

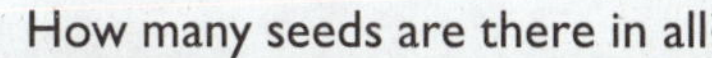

How many seeds are there in all? _____

There are 11 honeycombs and 10 cells in each one.

How many cells of honeycombs are there? _____

Fill in the Answer

Fill in the blanks to complete the equations.

10 × ___ = 120	12 × 10 = ___	12 × 10 = ___
10 × 12 = ___	___ × 12 = 120	___ × 10 = 120
___ × 12 = 120	10 × ___ = 120	12 × ___ = 120
12 × ___ = 120	10 × 12 = ___	___ × 12 = 120

---

10 × 12 = ___	___ × 10 = 120	12 × ___ = 120	___ × 10 = 120	12 × ___ = 120	12 × 10 = ___

---

12 × ___ = 120	10 × 12 = ___	___ × 10 = 120	12 × ___ = 120	10 × 12 = ___	___ × 10 = 120

Activities

There are ten light bulbs.

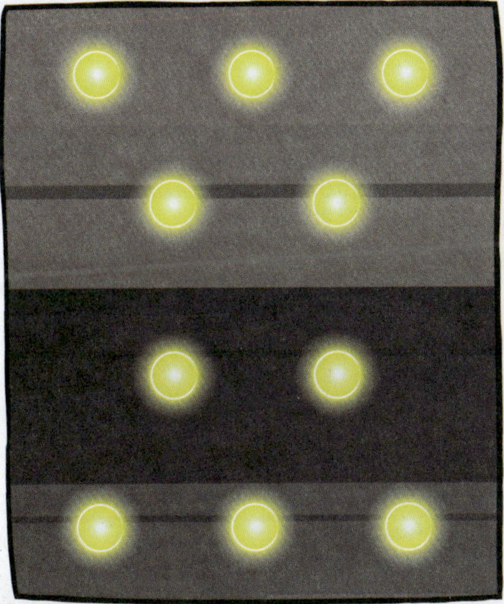

There are twelve fireflies around each light.

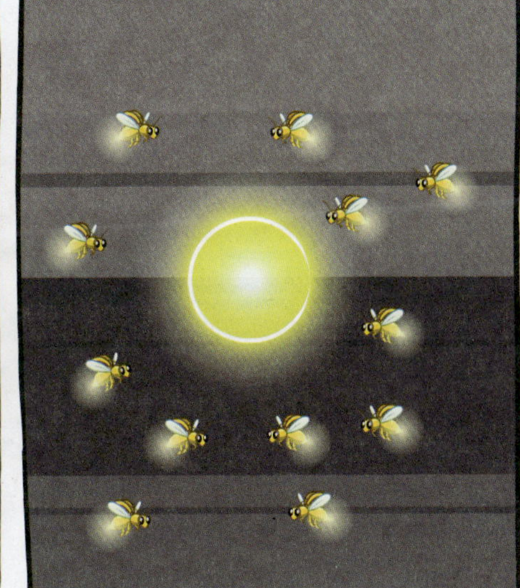

How many fireflies are there in all? _____

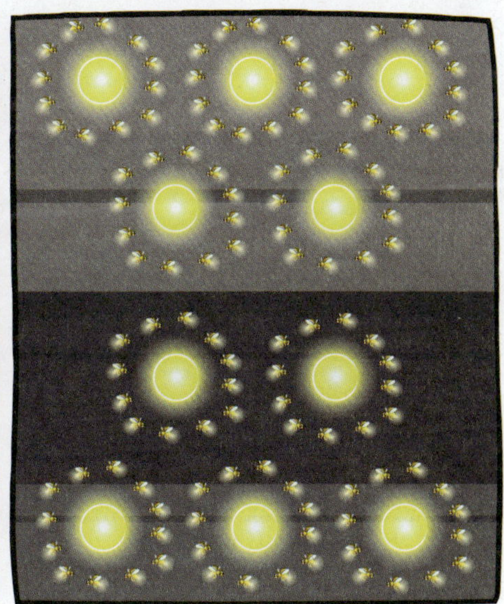

Use the numbers below to make equations.

There are 12 buttons. Each button has 10 holes.

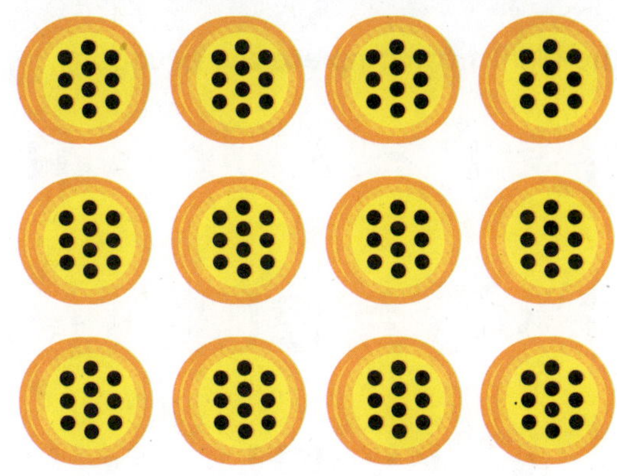

How many holes are there in all? _____

Draw ten candles on each holder.

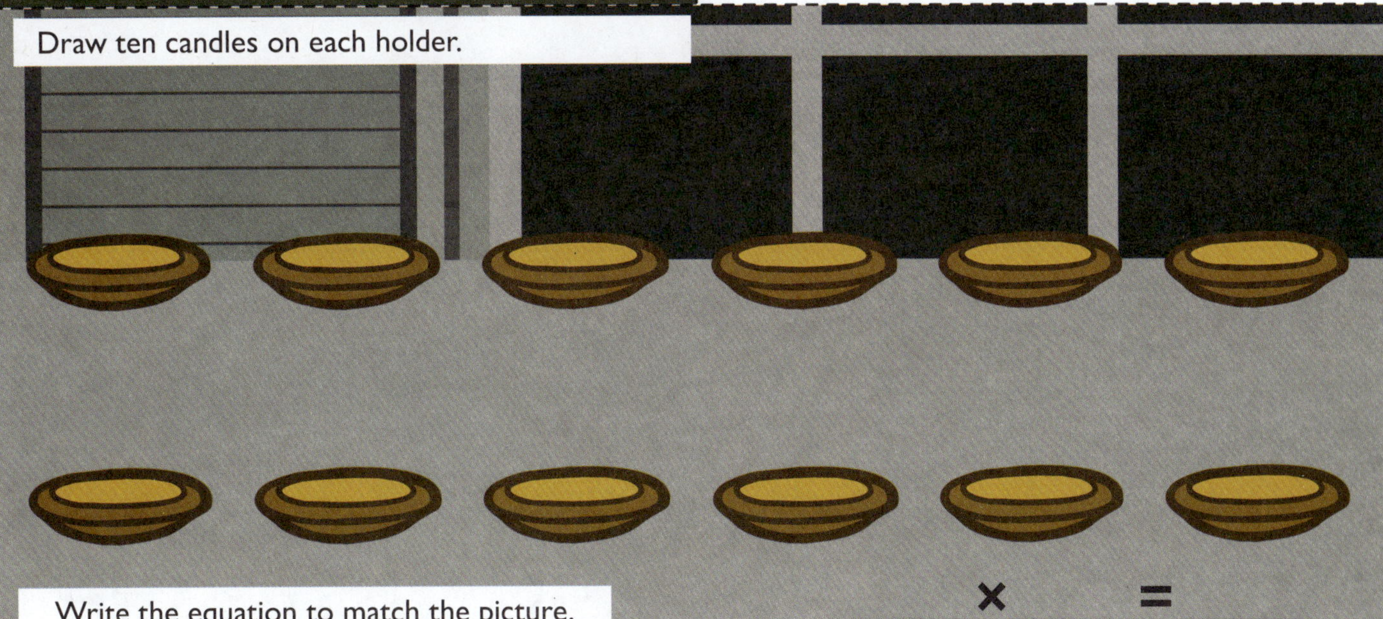

Write the equation to match the picture. _______ × _______ = _______

Fill in the Answer

Fill in the blanks to complete the equations.

11 × 11 = ___	11 × 11 = ___	___ × 11 = 121
11 × ___ = 121	___ × 11 = 121	11 × 11 = ___
___ × 11 = 121	11 × 11 = ___	___ × 11 = 121
11 × ___ = 121	11 × ___ = 121	11 × ___ = 121

11	___	11	___	11	11
× 11	× 11	× ___	× 11	× ___	× 11
___	121	121	121	121	___

11	11	___	11	11	___
× ___	× 11	× 11	× ___	× 11	× 11
121	___	121	121	___	121

Activities

There are eleven people. Each person has eleven flags. How many flags are there in all? _____

Use the numbers below to make equations.

11 11 121

11 × ___ = ___

___ × 11 = ___

11 × ___ = ___

___ × ___ = 121

Circle all of the 11, 11, 121 fact families.

11	11	121	5	11
11	9	5	11	35
121	5	121	30	121
11	11	121	11	30
3	11	11	121	1

Draw eleven flags on each hill.

Write the equation to match the picture.

_____ × _____ = _____

Fill in the Answer

Fill in the blanks to complete the equations.

$11 \times 12 = \square$ $11 \times \square = 132$ $\square \times 12 = 132$

$\square \times 12 = 132$ $\square \times 11 = 132$ $11 \times \square = 132$

$12 \times \square = 132$ $11 \times 12 = \square$ $12 \times 11 = \square$

$12 \times 11 = \square$ $12 \times \square = 132$ $\square \times 11 = 132$

$$\begin{array}{r} 11 \\ \times \ \square \\ \hline 132 \end{array} \quad \begin{array}{r} 11 \\ \times \ 12 \\ \hline \square \end{array} \quad \begin{array}{r} \square \\ \times \ 11 \\ \hline 132 \end{array} \quad \begin{array}{r} 12 \\ \times \ 11 \\ \hline \square \end{array} \quad \begin{array}{r} 11 \\ \times \ \square \\ \hline 132 \end{array} \quad \begin{array}{r} \square \\ \times \ 12 \\ \hline 132 \end{array}$$

$$\begin{array}{r} 12 \\ \times \ 11 \\ \hline \square \end{array} \quad \begin{array}{r} 12 \\ \times \ \square \\ \hline 132 \end{array} \quad \begin{array}{r} \square \\ \times \ 11 \\ \hline 132 \end{array} \quad \begin{array}{r} 12 \\ \times \ \square \\ \hline 132 \end{array} \quad \begin{array}{r} 11 \\ \times \ 12 \\ \hline \square \end{array} \quad \begin{array}{r} \square \\ \times \ 11 \\ \hline 132 \end{array}$$

Activities

There are eleven lines of rope.

Each line of rope has twelve lanterns.

How many lanterns are there in all? ____

Solve the crossword puzzle.

132		132		
=		=		
	×	11	=	132
×		×		
11	×		=	

Draw twelve lanterns on each building.

Write the equation to match the picture. ______ × ______ = ______

Fill in the Answer

Fill in the blanks to complete the equations.

$12 \times \square = 144$ $\square \times 12 = 144$ $\square \times 12 = 144$

$12 \times 12 = \square$ $12 \times 12 = \square$ $12 \times 12 = \square$

$\square \times 12 = 144$ $12 \times \square = 144$ $\square \times 12 = 144$

$12 \times \square = 144$ $12 \times 12 = \square$ $12 \times \square = 144$

$$\begin{array}{r} 12 \\ \times\ \square \\ \hline 144 \end{array} \quad \begin{array}{r} 12 \\ \times\ 12 \\ \hline \square \end{array} \quad \begin{array}{r} \square \\ \times\ 12 \\ \hline 144 \end{array} \quad \begin{array}{r} 12 \\ \times\ 12 \\ \hline \square \end{array} \quad \begin{array}{r} 12 \\ \times\ \square \\ \hline 144 \end{array} \quad \begin{array}{r} \square \\ \times\ 12 \\ \hline 144 \end{array}$$

$$\begin{array}{r} 12 \\ \times\ 12 \\ \hline \square \end{array} \quad \begin{array}{r} 12 \\ \times\ \square \\ \hline 144 \end{array} \quad \begin{array}{r} \square \\ \times\ 12 \\ \hline 144 \end{array} \quad \begin{array}{r} 12 \\ \times\ \square \\ \hline 144 \end{array} \quad \begin{array}{r} 12 \\ \times\ 12 \\ \hline \square \end{array} \quad \begin{array}{r} \square \\ \times\ 12 \\ \hline 144 \end{array}$$

Activities

There are twelve rocks. There are twelve starfish on each rock. How many starfish are there in all? ____

There are 12 ferns and each have 12 leaves.

How many leaves in all? _____

Draw twelve fins around each island.

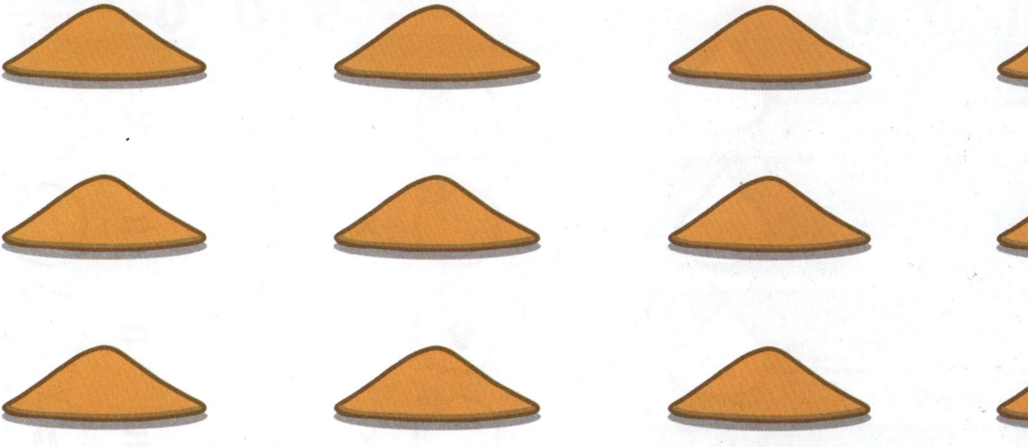

Write the equation to match the picture. _______ × _______ = _______

Zero Activities

Fill in the blanks to solve the equations.

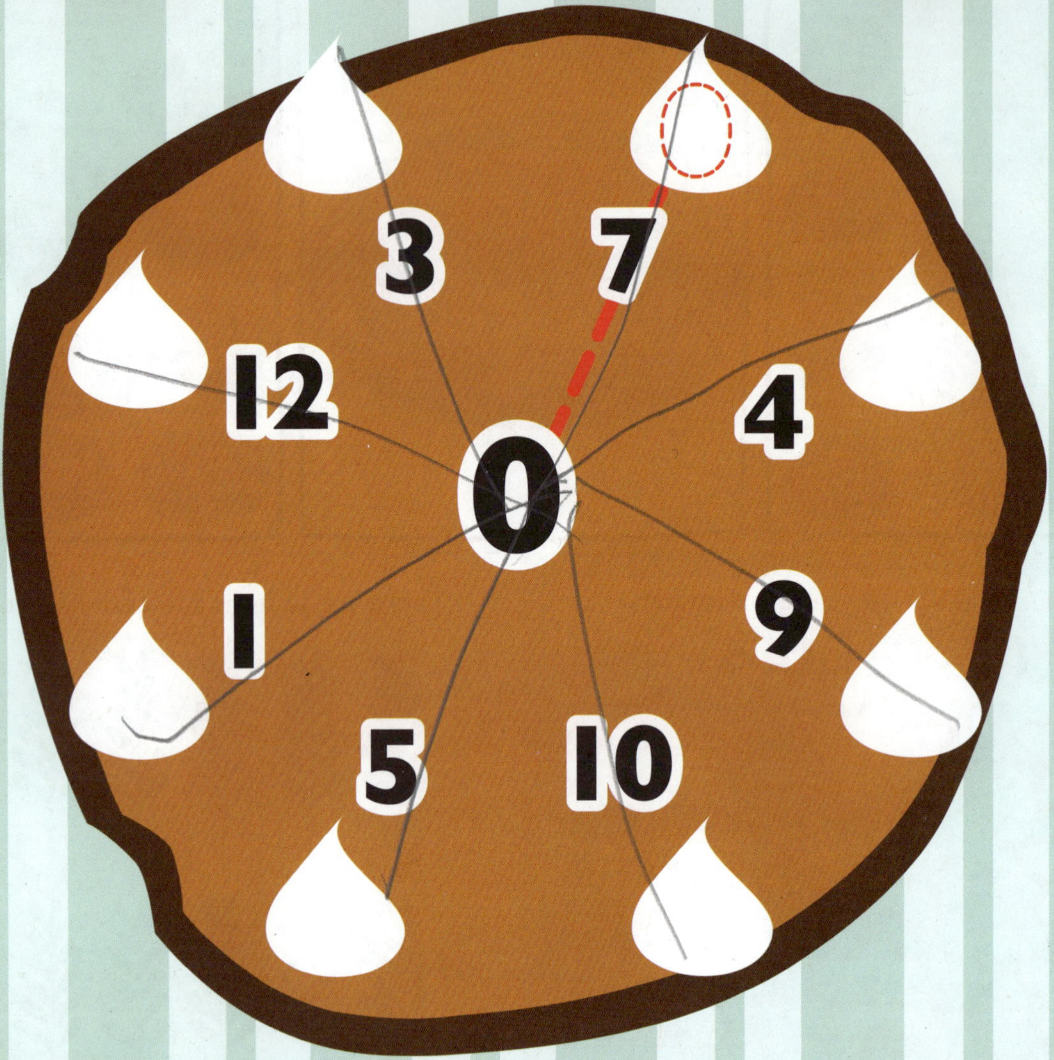

3 7 12 4 0 1 9 5 10

Zero Activities

There are five plates.

There are zero cookies on each plate.

How many cookies are there in all? ____

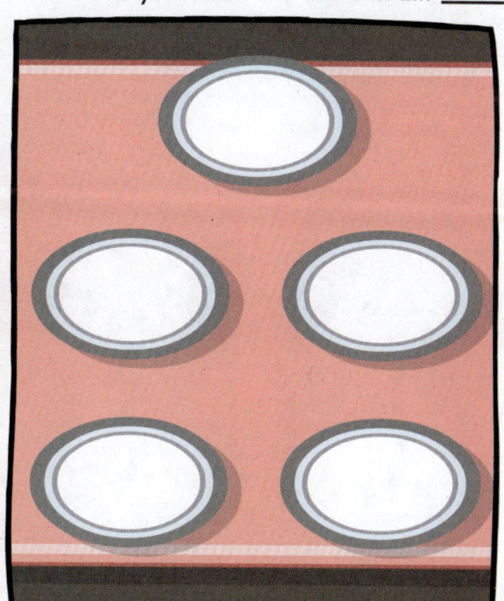

Use the numbers below to make equations.

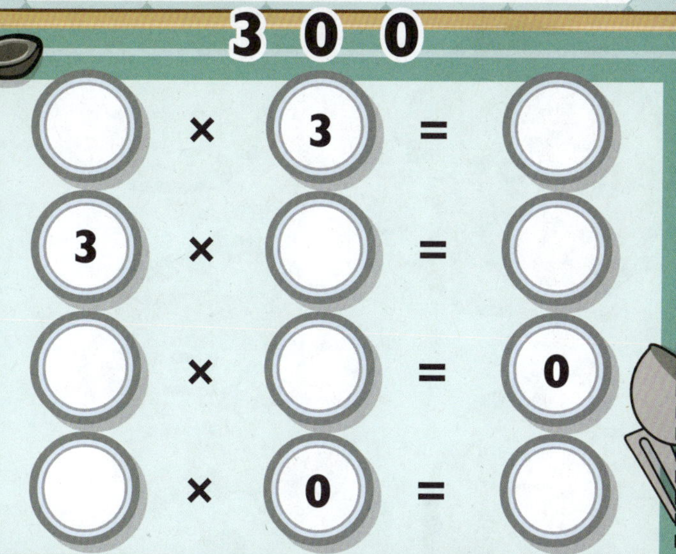

Complete the crossword puzzle.

4	×		=	0
×		×		
	×	8	=	0
=		=		
0	×	0	=	
				×
12		0		10
×		×		=
	×	6	=	
=		=		
	×		=	0

Solve the equations to get the cook to the bag of chocolate chips.

4×0= ☐

0×7= ☐

11×0= ☐

0×0= ☐

9×0= ☐

Equals 0

Equals 0

Fill in the blanks to solve the equations.
☐ ×7=0
0×11= ☐
1× ☐ =0
9× ☐ =0
☐ ×12=0
0×0= ☐
0×5= ☐
7× ☐ =0
☐ ×10=0

Skip Counting by 2

Skip Counting by 2

Skip Counting by 3

Skip Counting by 3

Skip Counting by 4

Use the number line to skip count by 4.

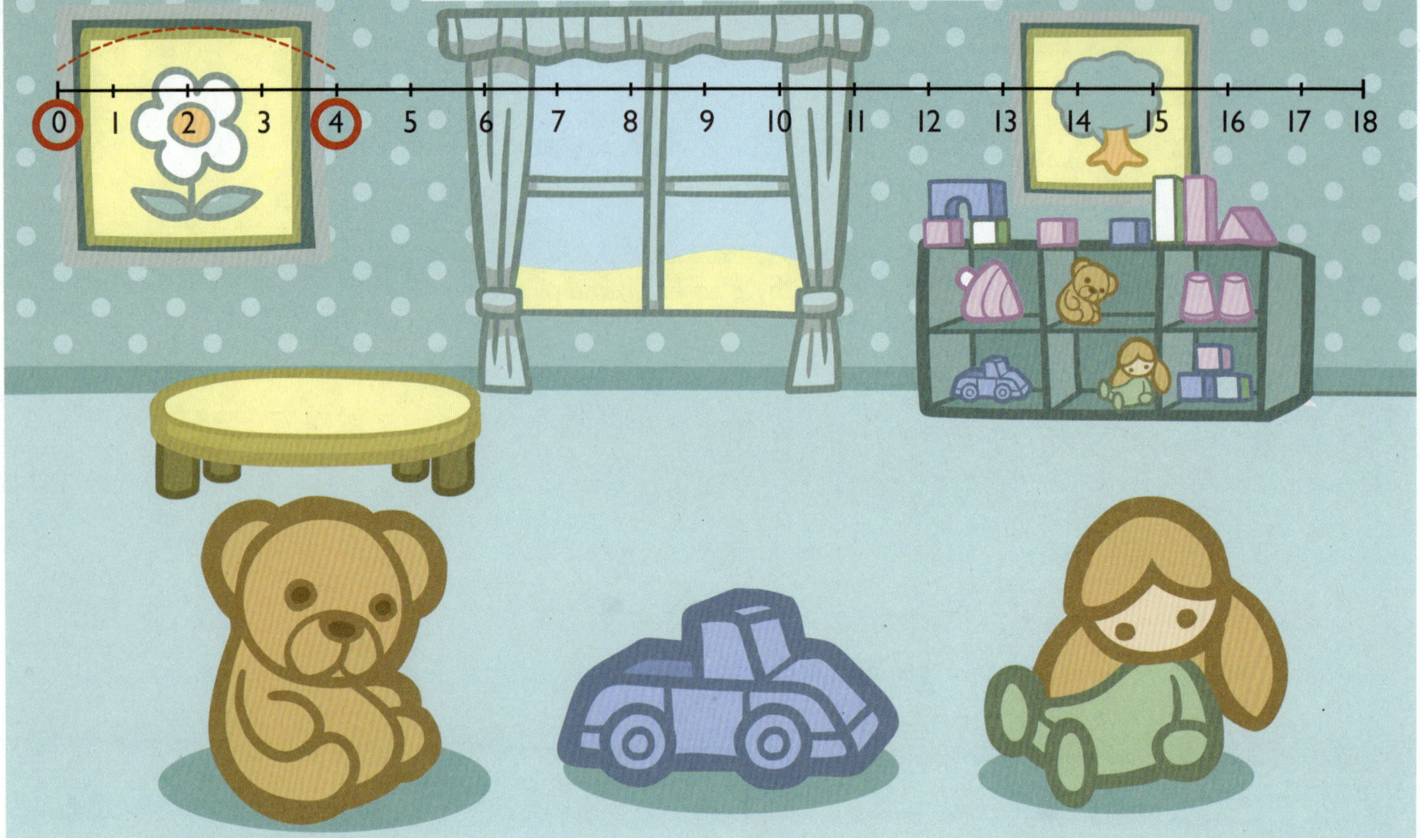

0 1 2 3 4 5 6 7 8 9 10 11 12 13 14 15 16 17 18

Skip counting by 4 to fill in the missing numbers.

Skip Counting by 4

Skip Counting by 5

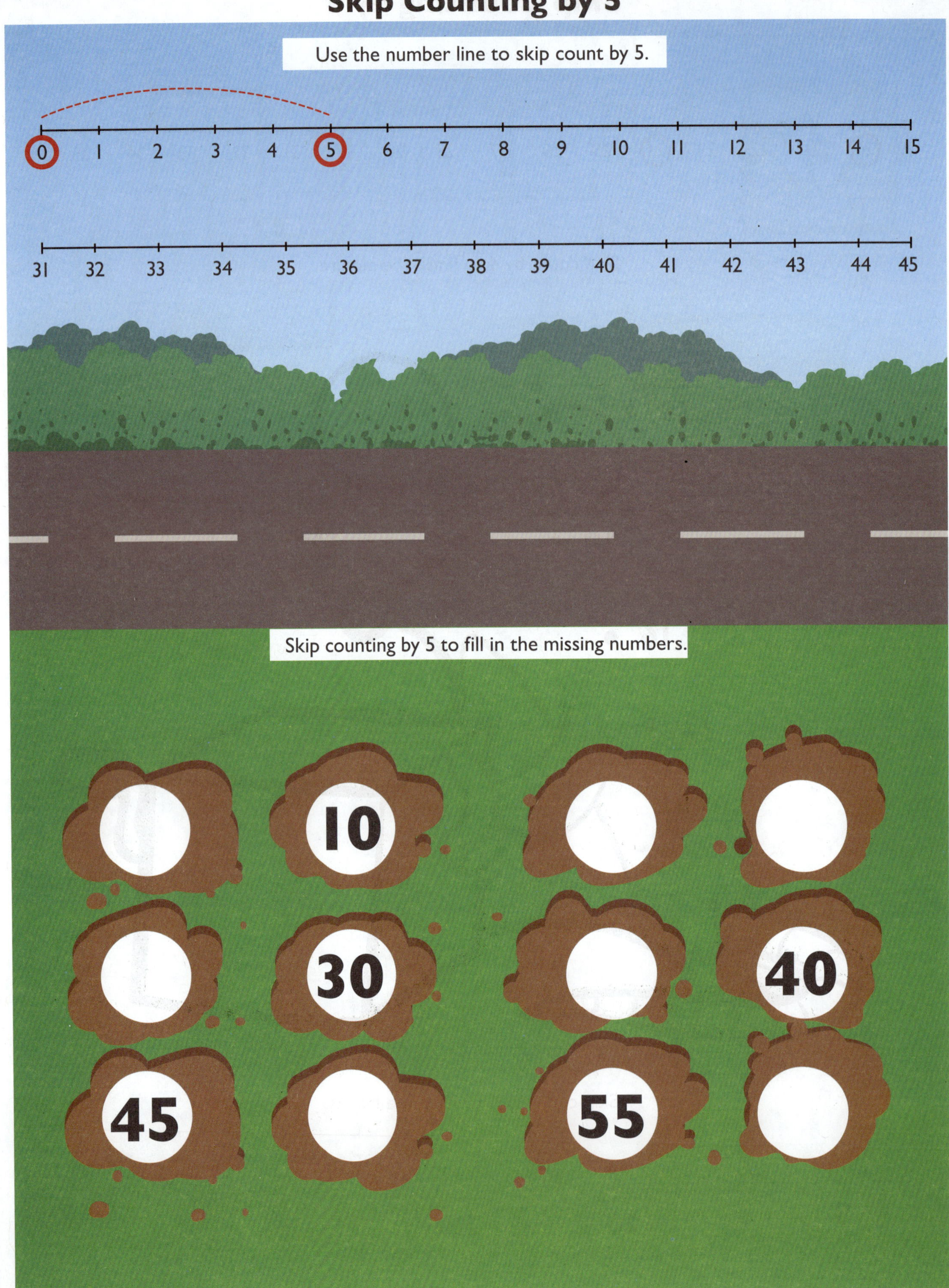

Skip Counting by 5

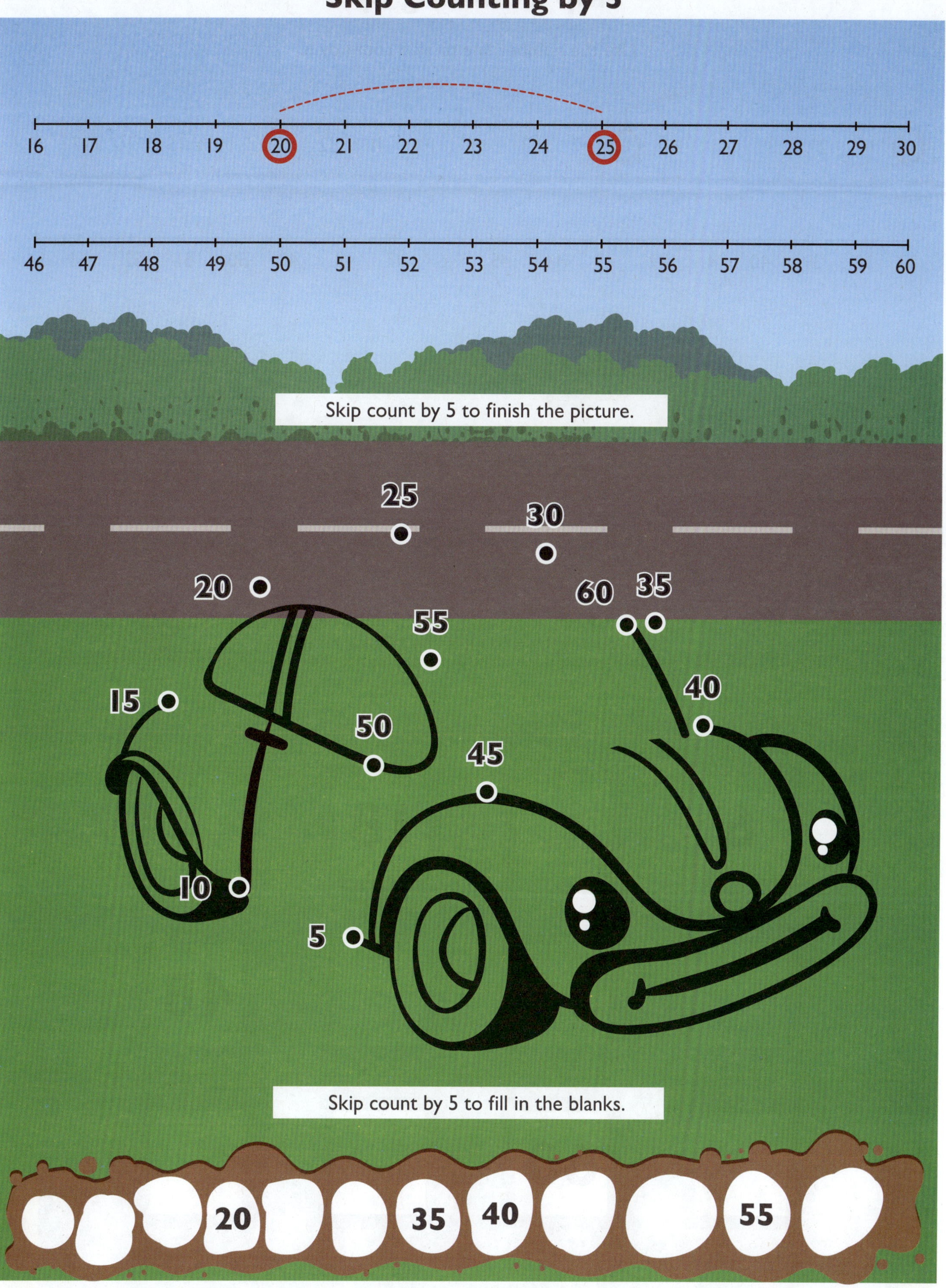

Skip Counting by 6

Skip Counting by 6

Skip Counting by 7

Skip Counting by 7

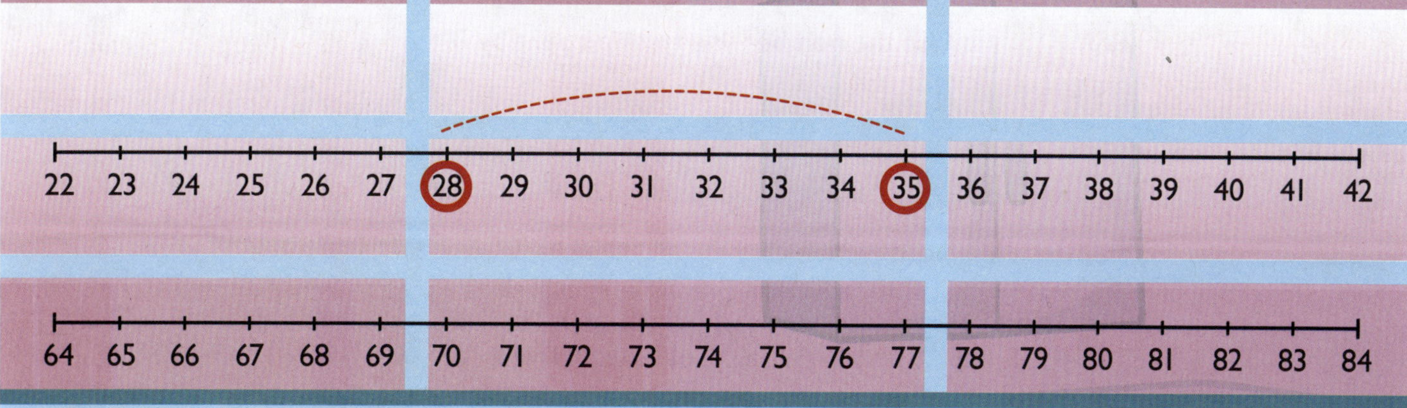

Skip count by 7 to finish the picture.

Skip count by 7 to fill in the blanks.

Skip Counting by 8

Skip Counting by 8

Skip Counting by 9

Skip Counting by 9

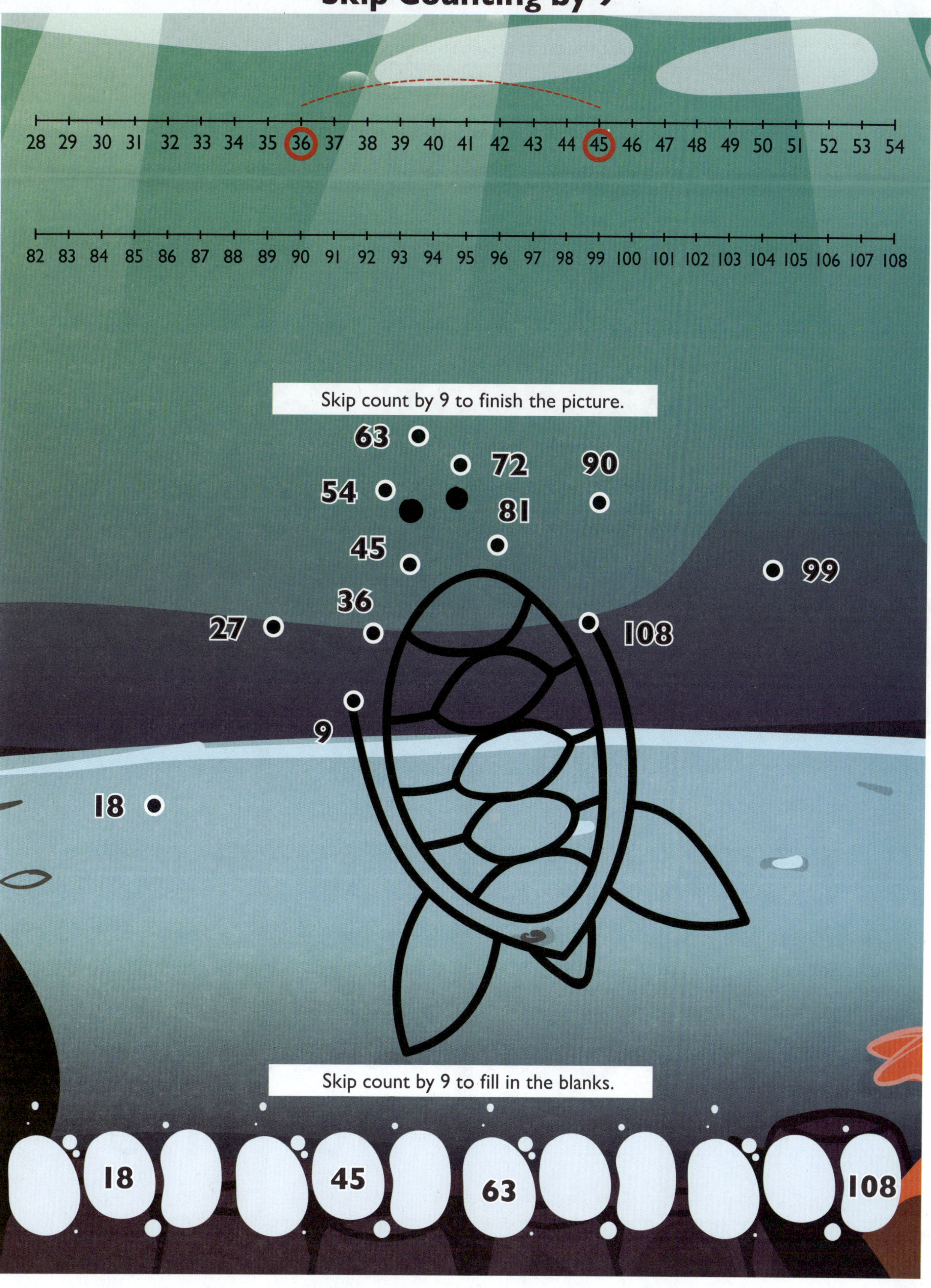

Skip Counting by 10

Skip Counting by 10

Skip Counting by 11

Skip Counting by 11

Skip Counting by 12

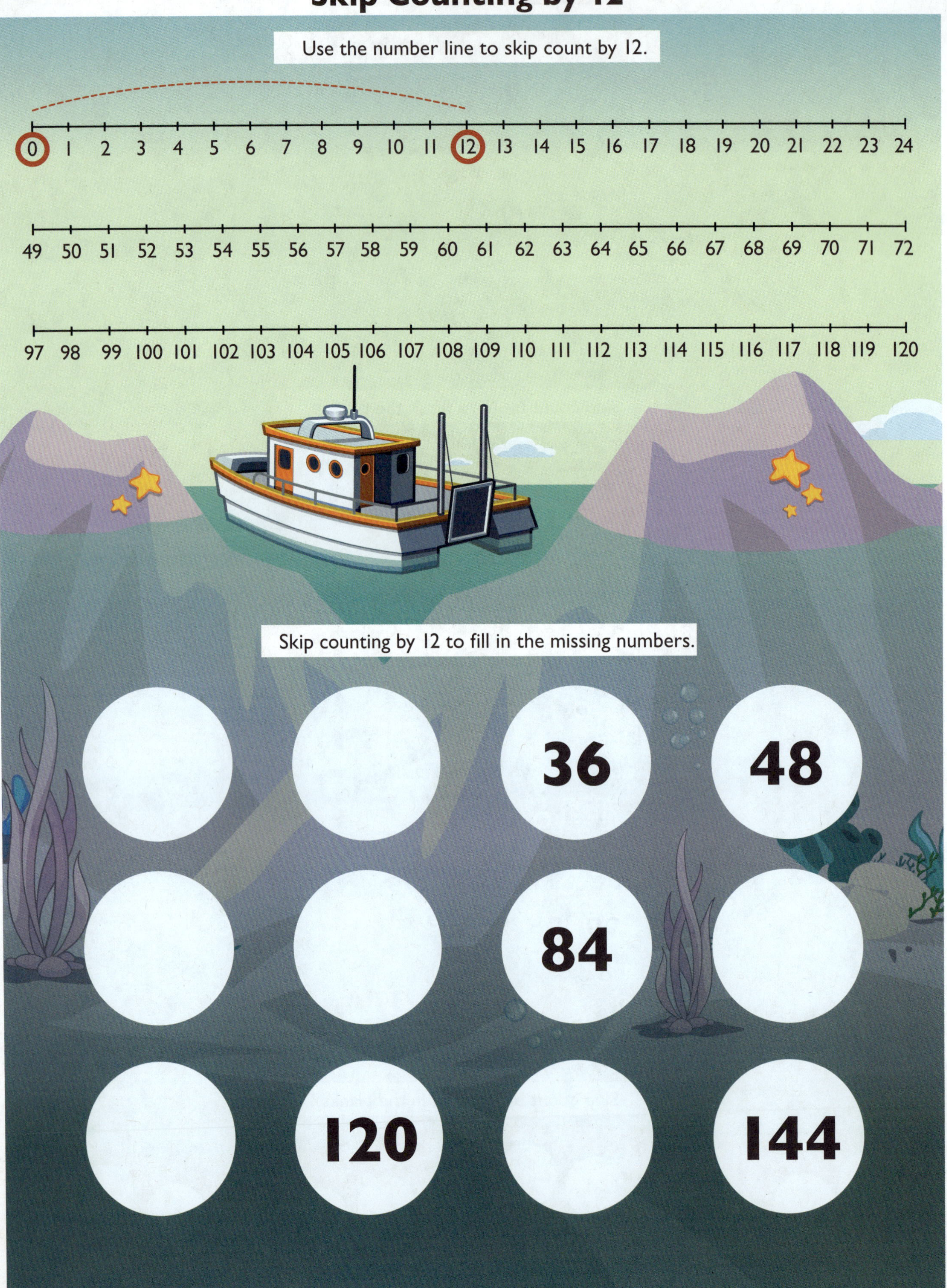

Skip Counting by 12

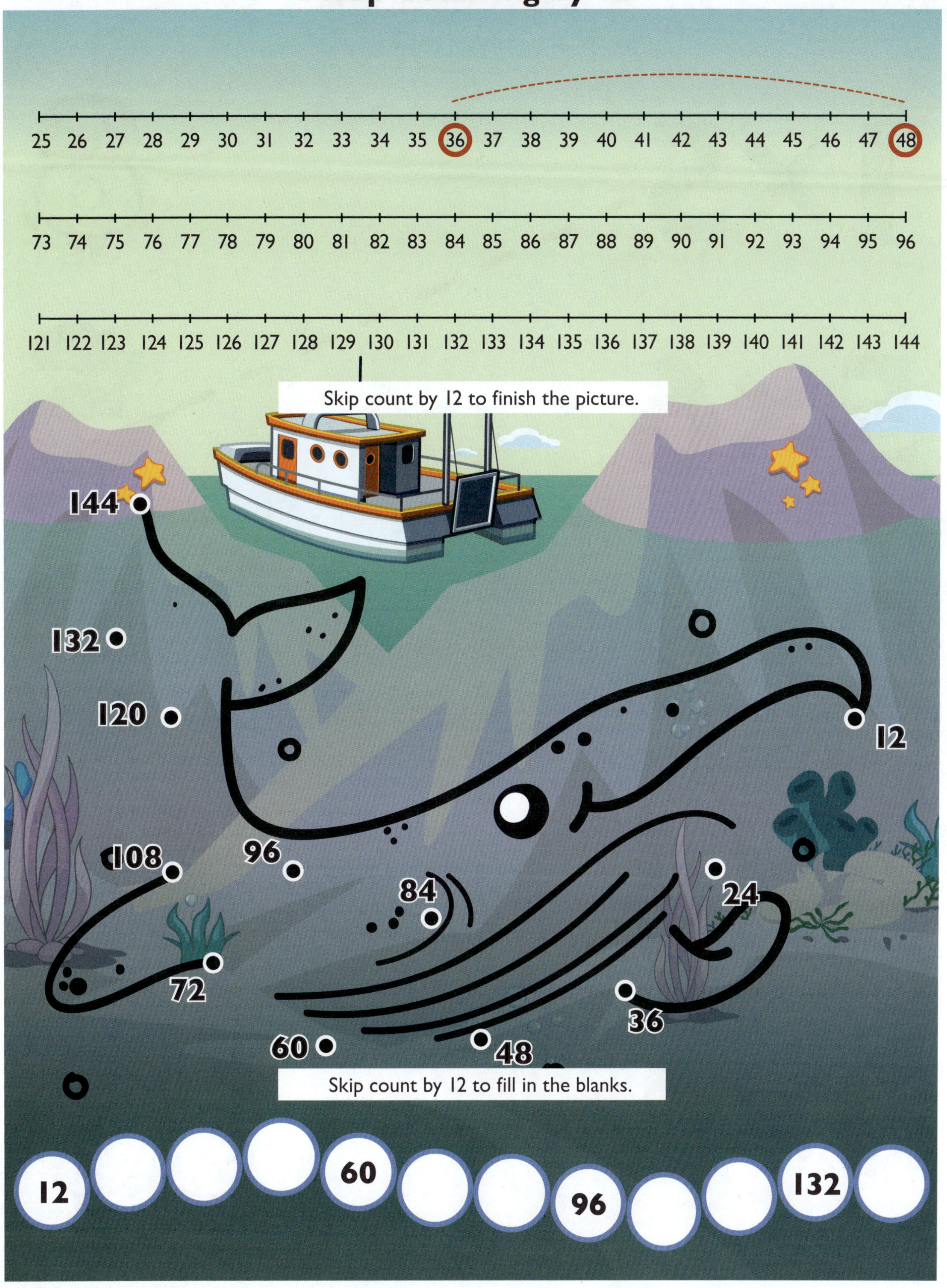

Multiplication Exercise

Connect the equations to the answers.

1×5

5×6

2×4

4×4

3×3

Multiplication Exercise

Connect the equations to the answers.

4×8

5×9

4×6

8×2

3×5

Multiplication Exercise

Connect the equations to the answers.

7×1

3×2

2×6

Multiplication Exercise

Connect the equations to the answers.

5×7

3×8

3×6

1×3

11×2

Multiplication Exercise

Connect the equations to the answers.

2×10

6×7

9×4

8×1

10×1

Multiplication Exercise

Connect the equations to the answers.

Multiplication Exercise

Connect the equations to the answers.

10×4

11×6

10×3

11×1

10×5

Multiplication Exercise

Connect the equations to the answers.

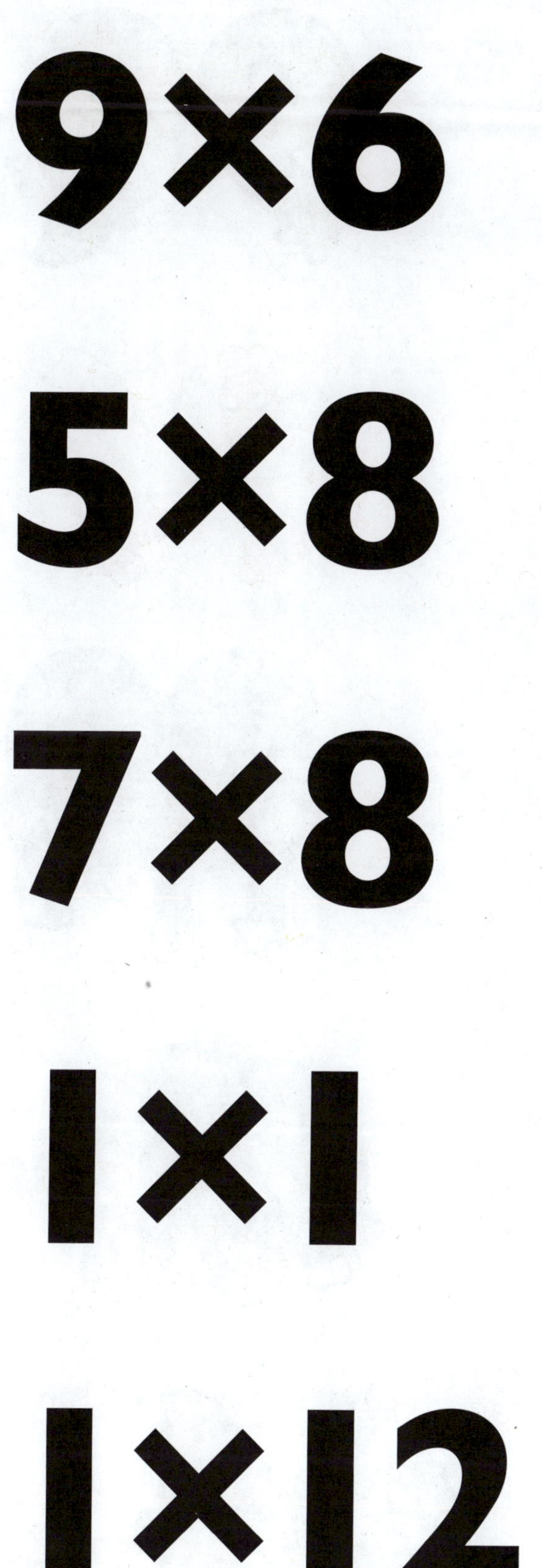

Multiplication Exercise

Connect the equations to the answers.

10×6

11×4

12×8

10×9

10×8

Multiplication Exercise

Connect the equations to the answers.

9×2	
2×5	
4×7	
6×8	
1×6	

Multiplication Exercise

Connect the equations to the answers.

2×12

9×12

5×12

3×11

8×11

Multiplication Exercise

Connect the equations to the answers.

1×2

3×4

5×5

6×6

3×9

Multiply by 1

Solve each problem and draw a line to connect bucket of paint to the easel.

Problem	Answer
7×1	1
1×1	7
6×1	10
8×1	6
10×1	8

Multiply by 1

Multiply the numbers by the center number and write the answer in the circle.

1×

3

5

9

2

4

4

Solve the equations to fill in the blanks with the corresponding letter.

A=3 T=5 R=6 F=4 S=9 I=2 U=8 N=10

1×3 1×6 1×5 1×2 1×9

Multiply by 2

Solve each problem and draw a line to connect the prospector to the bucket.

8×2

6×2

4×2

10×2

12×2

12

16

24

8

20

Multiply by 2

Multiply the numbers by the center number and write the answer in the circle.

Multiply by 2 to fill in the blanks.

×	9	5	7	6	4	10	12	8	0
2	18								

Multiply by 3

Solve each problem and draw a line to connect the carrot to the snowman.

Multiply by 3

Multiply the numbers by the center number and write the answer in the circle.

Solve the equations to fill in the blanks with the corresponding letter.

E=18 I=9 L=21 N=24 O=30 S=12 T=15 W=33

7×3 ____ 3×6 ____ 3×5 ____ 3×3 ____ 5×3 ____

3×4 ____ 3×8 ____ 3×10 ____ 3×11 ____

Multiply by 4

Solve each problem and draw a line to connect the lizard to the fly.

8×4	44
3×4	4
11×4	24
6×4	32
1×4	12

Multiply by 4

Multiply the numbers by the center number and write the answer in the circle.

Multiply by 4 to fill in the blanks.

×	3	8	11	9	4	5	12	7	10
4									

Multiply by 5

Solve each problem and draw a line to connect the top hat to the vest.

7×5	60
2×5	45
9×5	35
12×5	20
4×5	10

Multiply by 5

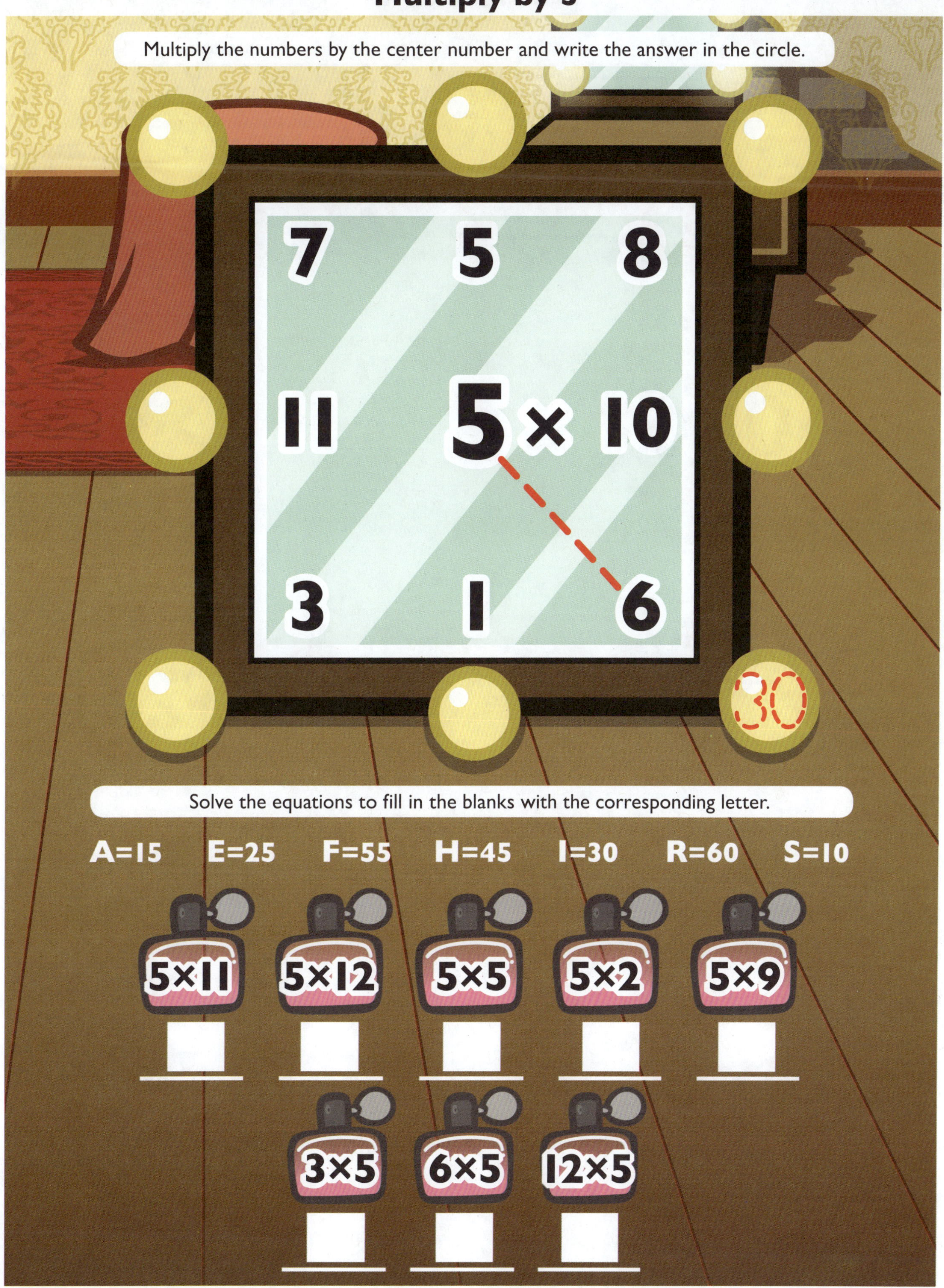

Multiply by 6

Solve each problem and draw a line to connect the arrow to the apple.

Problem	Answer
4×6	60
10×6	42
7×6	24
2×6	30
5×6	12

Multiply by 6

Multiply the numbers by the center number and write the answer in the circle.

48

8

9 6

12 11

6×

Multiply by 6 to fill in the blanks.

×	9	7	6	11	10	5	12	4	3
6									

Multiply by 7

Solve each problem and draw a line to connect the mallets to the xylophone.

1×7	49
11×7	28
7×7	7
9×7	77
4×7	63

Multiply by 7

Multiply the numbers by the center number and write the answer in the square.

2 — 14
10
12
7
7×
6
3
8
5

Solve the equations to fill in the blanks with the corresponding letter.

A=14 C=28 D=35 E=21 H=42 N=49 P=70 Y=63

Multiply by 8

Solve each problem and draw a line to connect the goggles to the backpack.

12×8

3×8

5×8

9×8

Multiply by 8

Multiply the numbers by the center number and write the answer in the circle.

8

1 8 4

8×

10 2 11

Multiply by 8 to fill in the blanks.

×	3	7	6	12	8	10	5	4	11
8									

Multiply by 9

Solve each problem and draw a line to connect the shovel to the hole.

Multiply by 9

Multiply the numbers by the center number and write the answer in the circle.

Solve the equations to fill in the blanks with the corresponding letter.

I=18 G=18 N=27 O=45 P=72 S=63 T=90

Multiply by 10

Solve each problem and draw a line to connect the airplane to the cone.

50

70

10

120

100

Multiply by 10

Multiply the numbers by the center number and write the answer in the square.

4 7

9 10 × 11

6 8

2 3

30

Multiply by 10 to fill in the blanks.

×	5	8	6	12	9	10	3	2	11
10									

Multiply by 11

Solve each problem and draw a line to connect the dolphin to the fish.

Dolphin	Fish
8×11	33
3×11	55
12×11	88
5×11	66
6×11	132

Multiply by 11

Multiply the numbers by the center number and write the answer in the triangle.

11×

2

11 7 77

9 6

Solve the equations to fill in the blanks with the corresponding letter.

A=22 F=55 H=33 I=99 R=44 S=110 T=77

10×11 ___ 7×11 ___ 2×11 ___ 4×11 ___

11×5 ___ 9×11 ___ 11×10 ___ 11×3 ___

Multiply by 12

Solve each problem and draw a line to connect the carrot to the rabbit.

4×12

24

48

120

84

108

Multiply by 12

Multiply the numbers by the center number and write the answer in the circle.

12 ×

3 → 36

5

11

8

1

6

Multiply by 12 to fill in the blanks.

×	5	8	6	12	9	10	3	2	11
12									

Test Your Skills Multiply By 0

$4 \times 0 =$ ____ $0 \times 5 =$ ____ $0 \times 3 =$ ____

$7 \times 0 =$ ____ $0 \times 2 =$ ____ $6 \times 0 =$ ____

$0 \times 4 =$ ____ $1 \times 0 =$ ____ $2 \times 0 =$ ____

$0 \times 8 =$ ____ $5 \times 0 =$ ____ $10 \times 0 =$ ____

$0 \times 6 =$ ____ $3 \times 0 =$ ____ $0 \times 7 =$ ____

$0 \times 9 =$ ____ $12 \times 0 =$ ____ $8 \times 0 =$ ____

$11 \times 0 =$ ____ $9 \times 0 =$ ____

Test Your Skills Multiply By 1

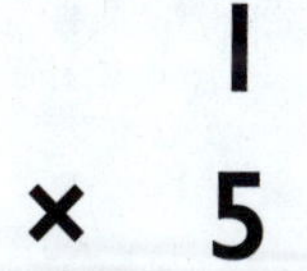

1×5	8×1	5×1	9×1	
6×1	1×2	1×9	1×8	3×1
7×1	1×3	4×1	10×1	1×11
1×7	0×1	1×1	11×1	12×1
1×6	7×1	1×10	8×1	1×0
9×1	1×5	1×12	1×4	2×1

Test Your Skills Multiply By 2

2	3	0	2	2	1
× 2	× 2	× 2	× 4	× 8	× 2
8	2	5	9	2	4
× 2	× 3	× 2	× 2	× 1	× 2
7	10	2	2	2	2
× 2	× 2	× 6	× 5	× 2	× 7
2	11	12	2	2	2
× 10	× 2	× 2	× 9	× 11	× 12
3	2	2	9	1	2
× 2	× 0	× 6	× 2	× 2	× 8
2	4	6	12		
× 10	× 2	× 2	× 2		

Test Your Skills Multiply By 3

1×3 5×3 9×3 3×5 2×3

6×3 8×3 3×1 10×3 3×3 3×2

7×3 4×3 3×3 3×6 3×9 3×0

3×7 3×8 11×3 3×12 3×11 12×3

9×3 3×10 5×3 3×3 3×8 7×3

12×3 3×2 4×3 3×6 0×3 3×1

Test Your Skills Multiply By 4

4 × 5 = ____ 4 × 3 = ____

2 × 4 = ____ 7 × 4 = ____ 4 × 10 = ____

5 × 4 = ____ 4 × 12 = ____ 4 × 7 = ____

10 × 4 = ____ 4 × 9 = ____ 4 × 11 = ____

1 × 4 = ____ 4 × 8 = ____ 4 × 6 = ____

9 × 4 = ____ 4 × 2 = ____ 4 × 4 = ____

0 × 4 = ____ 4 × 4 = ____ 11 × 4 = ____

Test Your Skills Multiply By 5

3 × 5 = ____ 10 × 5 = ____

5 × 5 = ____ 8 × 5 = ____ 5 × 7 = ____

5 × 0 = ____ 5 × 11 = ____ 2 × 5 = ____

5 × 2 = ____ 9 × 5 = ____ 12 × 5 = ____

5 × 1 = ____ 1 × 5 = ____ 7 × 5 = ____

5 × 9 = ____ 6 × 5 = ____ 4 × 5 = ____

5 × 12 = ____ 5 × 10 = ____ 11 × 5 = ____

Test Your Skills Multiply By 6

$2 \times 6 =$ ____ $5 \times 6 =$ ____

$9 \times 6 =$ ____ $6 \times 7 =$ ____ $7 \times 6 =$ ____

$1 \times 6 =$ ____ $6 \times 2 =$ ____ $6 \times 6 =$ ____

$4 \times 6 =$ ____ $6 \times 10 =$ ____ $11 \times 6 =$ ____

$3 \times 6 =$ ____ $6 \times 0 =$ ____ $12 \times 6 =$ ____

$10 \times 6 =$ ____ $6 \times 12 =$ ____ $6 \times 5 =$ ____

$8 \times 6 =$ ____ $6 \times 9 =$ ____ $6 \times 11 =$ ____

Test Your Skills Multiply By 7

	7×7	7×1	6×7	7×3	12×7
7×10	9×7	7×6	11×7	7×2	3×7
7×5	8×7	7×7	1×7	7×8	0×7
7×9	10×7	7×11	4×7	7×12	5×7
7×4	5×7	2×7	7×10	7×9	4×7
7×0	3×7	7×11	6×7	7×8	7×7

Test Your Skills Multiply By 8

6	8	10	8	3	8
× 8	× 9	× 8	× 5	× 8	× 7
1	8	8	8	7	8
× 8	× 0	× 8	× 12	× 8	× 4
5	8	4	8	2	8
× 8	× 3	× 8	× 6	× 8	× 8
11	8	12	8	9	8
× 8	× 10	× 8	× 11	× 8	× 2
4	8	8	7	6	0
× 8	× 1	× 5	× 8	× 8	× 8
8	2	8	8	3	
× 9	× 8	× 10	× 12	× 8	

Test Your Skills Multiply By 9

9 × 0 = ____ 3 × 9 = ____

8 × 9 = ____ 9 × 3 = ____ 1 × 9 = ____

9 × 2 = ____ 6 × 9 = ____ 9 × 10 = ____

4 × 9 = ____ 9 × 4 = ____ 2 × 9 = ____

9 × 7 = ____ 5 × 9 = ____ 9 × 11 = ____

12 × 9 = ____ 9 × 8 = ____ 7 × 9 = ____

9 × 6 = ____ 10 × 9 = ____ 9 × 9 = ____

Test Your Skills Multiply By 10

10 × 5 = ____ 6 × 10 = ____

10 × 8 = ____ 9 × 10 = ____ 10 × 6 = ____

3 × 10 = ____ 10 × 4 = ____ 7 × 10 = ____

10 × 7 = ____ 2 × 10 = ____ 10 × 2 = ____

4 × 10 = ____ 10 × 9 = ____ 8 × 10 = ____

10 × 1 = ____ 5 × 10 = ____ 10 × 3 = ____

2 × 10 = ____ 10 × 11 = ____ 0 × 10 = ____

Test Your Skills Multiply By 11

11 × 4 = ____ 6 × 11 = ____ 11 × 3 = ____

8 × 11 = ____ 11 × 8 = ____ 2 × 11 = ____

11 × 6 = ____ 7 × 11 = ____ 11 × 9 = ____

4 × 11 = ____ 11 × 5 = ____ 3 × 11 = ____

11 × 0 = ____ 9 × 11 = ____ 11 × 9 = ____

5 × 11 = ____ 11 × 1 = ____ 11 × 11 = ____

11 × 7 = ____ 12 × 11 = ____

Test Your Skills Multiply By 12

12 × 1 = ____	8 × 12 = ____	12 × 3 = ____
7 × 12 = ____	12 × 0 = ____	6 × 12 = ____
12 × 2 = ____	1 × 12 = ____	12 × 9 = ____
9 × 12 = ____	12 × 7 = ____	11 × 12 = ____
12 × 6 = ____	4 × 12 = ____	12 × 7 = ____
2 × 12 = ____	12 × 4 = ____	5 × 12 = ____
	12 × 12 = ____	10 × 12 = ____

Mixed Reviews

1×5 | 8×10 | 9×8 | 6×5

6×3 | 7×2 | 1×4 | 8×5 | 3×5

0×1 | 8×3 | 5×5 | 8×7 | 6×2

4×7 | 6×9 | 1×1 | 6×4 | 4×2

5×0 | 7×5 | 2×3 | 7×7 | 2×2

3×3 | 6×6 | 12×2 | 6×12 | 11×6

Mixed Reviews

$11 \times 5 =$ ____ $7 \times 10 =$ ____ $1 \times 10 =$ ____

$7 \times 5 =$ ____ $10 \times 2 =$ ____ $6 \times 9 =$ ____

$12 \times 2 =$ ____ $1 \times 10 =$ ____ $6 \times 4 =$ ____

$9 \times 9 =$ ____ $7 \times 7 =$ ____ $10 \times 12 =$ ____

$5 \times 6 =$ ____ $4 \times 12 =$ ____ $11 \times 7 =$ ____

$5 \times 12 =$ ____ $5 \times 4 =$ ____ $0 \times 8 =$ ____

$4 \times 4 =$ ____ $10 \times 10 =$ ____

Mixed Reviews

$$\begin{array}{r} 9 \\ \times\ 5 \\ \hline \end{array} \quad \begin{array}{r} 5 \\ \times\ 11 \\ \hline \end{array} \quad \begin{array}{r} 9 \\ \times\ 8 \\ \hline \end{array} \quad \begin{array}{r} 2 \\ \times\ 5 \\ \hline \end{array}$$

$$\begin{array}{r} 8 \\ \times\ 4 \\ \hline \end{array} \quad \begin{array}{r} 2 \\ \times\ 2 \\ \hline \end{array} \quad \begin{array}{r} 9 \\ \times\ 2 \\ \hline \end{array} \quad \begin{array}{r} 3 \\ \times\ 7 \\ \hline \end{array} \quad \begin{array}{r} 1 \\ \times\ 8 \\ \hline \end{array}$$

$$\begin{array}{r} 3 \\ \times\ 9 \\ \hline \end{array} \quad \begin{array}{r} 7 \\ \times\ 4 \\ \hline \end{array} \quad \begin{array}{r} 0 \\ \times\ 11 \\ \hline \end{array} \quad \begin{array}{r} 6 \\ \times\ 2 \\ \hline \end{array} \quad \begin{array}{r} 4 \\ \times\ 2 \\ \hline \end{array}$$

$$\begin{array}{r} 1 \\ \times\ 3 \\ \hline \end{array} \quad \begin{array}{r} 5 \\ \times\ 12 \\ \hline \end{array} \quad \begin{array}{r} 1 \\ \times\ 10 \\ \hline \end{array} \quad \begin{array}{r} 4 \\ \times\ 12 \\ \hline \end{array} \quad \begin{array}{r} 10 \\ \times\ 2 \\ \hline \end{array}$$

$$\begin{array}{r} 7 \\ \times\ 5 \\ \hline \end{array} \quad \begin{array}{r} 8 \\ \times\ 2 \\ \hline \end{array} \quad \begin{array}{r} 9 \\ \times\ 12 \\ \hline \end{array} \quad \begin{array}{r} 8 \\ \times\ 8 \\ \hline \end{array} \quad \begin{array}{r} 2 \\ \times\ 3 \\ \hline \end{array}$$

$$\begin{array}{r} 9 \\ \times\ 9 \\ \hline \end{array} \quad \begin{array}{r} 7 \\ \times\ 6 \\ \hline \end{array} \quad \begin{array}{r} 12 \\ \times\ 5 \\ \hline \end{array} \quad \begin{array}{r} 3 \\ \times\ 11 \\ \hline \end{array} \quad \begin{array}{r} 10 \\ \times\ 10 \\ \hline \end{array}$$